914.70~~3~~ ~~~~
SXM   SYM

2.8.76

RUSSIAN TRANSPORT

An historical and geographical survey

BELL'S ADVANCED ECONOMIC GEOGRAPHIES

General Editor
PROFESSOR R. O. BUCHANAN
M.A.(N.Z.), B.Sc.(Econ.), Ph.D.(London)
Professor Emeritus, University of London

A. Systematic Studies

PLANTATION AGRICULTURE
P. P. Courtenay, B.A., Ph.D.

NEW ENGLAND: A STUDY IN INDUSTRIAL ADJUSTMENT
R. C. Estall, B.Sc.(Econ.), Ph.D.

GEOGRAPHY AND ECONOMICS
Michael Chisholm, M.A.

AGRICULTURAL GEOGRAPHY
Leslie Symons, B.Sc.(Econ.), Ph.D.

REGIONAL ANALYSIS AND ECONOMIC GEOGRAPHY
John N. H. Britton, M.A., Ph.D.

THE FISHERIES OF EUROPE: AN ECONOMIC GEOGRAPHY
James R. Coull, M.A., Ph.D.

A GEOGRAPHY OF TRADE AND DEVELOPMENT IN MALAYA
P. P. Courtenay, B.A., Ph.D.

R. O. BUCHANAN AND ECONOMIC GEOGRAPHY
(Ed.) *M. J. Wise, M.C., B.A., Ph.D. & E. M. Rawstron, M.A.*

B. Regional Studies

AN ECONOMIC GEOGRAPHY OF EAST AFRICA
A. M. O'Connor, B.A., Ph.D.

AN ECONOMIC GEOGRAPHY OF WEST AFRICA
H. P. White, M.A. & M. B. Gleave, M.A.

AN ECONOMIC GEOGRAPHY OF ROMANIA
David Turnock, M.A., Ph.D.

YUGOSLAVIA: PATTERNS OF ECONOMIC ACTIVITY
F. E. Ian Hamilton, B.Sc.(Econ.), Ph.D.

RUSSIAN AGRICULTURE: A GEOGRAPHIC SURVEY
Leslie Symons, B.Sc.(Econ.), Ph.D.

RUSSIAN TRANSPORT
(Ed.) *Leslie Symons, B.Sc.(Econ.), Ph.D. & Colin White, B.A.(Cantab.)*

AN AGRICULTURAL GEOGRAPHY OF GREAT BRITAIN
J. T. Coppock, M.A., Ph.D.

AN HISTORICAL INTRODUCTION TO THE ECONOMIC GEOGRAPHY OF GREAT BRITAIN
Wilfred Smith, M.A.

THE BRITISH IRON & STEEL SHEET INDUSTRY SINCE 1840
Kenneth Warren, M.A., Ph.D.

A GEOGRAPHY OF BRAZILIAN DEVELOPMENT
Janet D. Henshall, M.A., M.Sc., Ph.D. & R. P. Momsen Jr, A.B., M.A., Ph.D.

RUSSIAN TRANSPORT

An historical and geographical survey

Edited by

Leslie Symons, B.Sc. (Econ.), Ph.D.
Reader in the Geography of Russia
University College, Swansea

and

Colin White, B.A. (Cantab.)
Lecturer in Russian Economic History
University College, Swansea

LONDON

G. BELL & SONS LTD

1975

Copyright © 1975 by
G. BELL AND SONS, LTD
York House, Portugal Street
London, WC2A 2HL

First published 1975

All rights reserved. No part of this publication
may be reproduced, stored in a retrieval system,
or transmitted, in any form or by any means,
electronic, mechanical, photocopying, recording
or otherwise, without the prior permission of
G. Bell and Sons, Ltd.

India
    Orient Longman Ltd, Calcutta, Bombay, Madras and New Delhi
Canada
    Clarke, Irwin & Co. Ltd, Toronto
Australia
    Edward Arnold (Australia) Pty Ltd, Port Melbourne, Vic.
New Zealand
    Book Reps (New Zealand) Ltd, 46 Lake Road, Northcote, Auckland
East Africa
    J. E. Budds, P.O. Box 4536, Nairobi
West Africa
    Thos. Nelson (Nigeria) Ltd, P.O. Box 336, Apapa, Lagos
South and Central Africa
    Book Promotions (Pty) Ltd, 311 Sanlam Centre, Main Road
    Wynberg, Cape Province

ISBN Hardback 0 7135 1895 2
ISBN Paperback 0 7135 1911 8

*Printed in Great Britain by
The Camelot Press Ltd, Southampton*

Preface

This book had its origins in a colloquium 'The role of transport in Russian economic development' held at Gregynog Hall, the conference centre of the University of Wales near Newtown, Montgomeryshire, 19–21 September 1972. The colloquium was organised by the Centre of Russian and East European Studies, University College of Swansea, and financed by the College. The subject of the colloquium was so chosen because it appeared to the organisers that relatively little attention had been paid by specialists in Russian and Soviet studies to the various aspects of transport, whether from an historical or geographical point of view, and it seemed to be an excellent subject for demonstrating the advantages of interdisciplinary links in regional studies. The organisers were fortunate in being able to secure the services of several of the small number of workers in the field in Britain to write papers specially for the colloquium, and to present these papers to those who were able to take the opportunity to attend the gathering at Gregynog. Ample time was reserved for discussion of papers but it was decided not to publish the many contributions made in this way because of the problems of recording, editing and cost.

The Centre of Russian and East European Studies is indebted to Dr Glyn Tegai Hughes, Warden of Gregynog Hall, and his staff for their hospitality at this delightful country estate bequeathed to the University of Wales by Miss Margaret Davies, to Professor R. O. Buchanan for reading the manuscript of the book and making helpful suggestions, to Mr K. Vomhof for similar help with Chapter 6, and to Mr G. B.

Lewis, Department of Geography, University College of Swansea, for the cartography. The authors of the various papers are solely responsible for any opinions or views which may appear in the corresponding chapters, the editors being similarly responsible for the introductory and concluding chapters.

It is hoped that the Centre of Russian and East European Studies at Swansea will be responsible for further volumes of a similar kind in the future.

Swansea
Summer 1974

L. J. S.
C. W.

Contents

			Page
	Maps and diagrams		viii
	Tables		ix
	Contributors		xi
	Note on transliteration		xiii
	Weights and measures		xiv
	Introduction	The editors	xv
1	The impact of Russian railway construction on the market for grain in the 1860s and 1870s	C. White	1
2	Railways and economic development in Turkestan before 1917	D. Spring	46
3	The Soviet concept of a unified transport system and the contemporary role of the railways	R. E. H. Mellor	75
4	The Soviet merchant marine	R. H. Greenwood	106
5	The northern sea route	T. Armstrong	127
6	Soviet air transport	L. Symons	142
7	Conclusions	The editors	164
	Some important dates in the history of transport in Russia and the U.S.S.R.		179
	Index		183

Maps and Diagrams

		Page
1	Russia's railway network in 1875	3
2	The provinces of tsarist Russia	15
3	Grain supply routes of Moscow	31
4	Grain supply routes of St Petersburg	35
5	Railways in Russian Central Asia before 1917	48
6	Kazanskiy's concept of the unified transport system and belts of major economic activity	81
7	Main interregional freight flows (modified from Kazanskiy)	82
8	Freight flows of selected commodities by railway	84–5
9	The unified transport system—an essay	101
10	The major regular cargo-liner services operated by the U.S.S.R. and the approximate areas served, 1972	120
11	The major Comecon joint cargo-liner routes and other routes served by Polish Ocean Lines, 1972	121
12	The northern sea route and principal transport routes in Siberia and the Soviet Far East	128
13	Aeroflot services in 1936	145
14	The Aeroflot international network in 1972	152
15	Internal air routes of major interregional significance operating in 1972	154
16	The Aeroflot network in part of western Siberia	156
17	The growth of total freight traffic movement and movement by each of the main media in the U.S.S.R., 1928–71	168
18	The growth of total passenger traffic movement and movement by each of the main media in the U.S.S.R., 1928–71	173

Tables

		Page
1	Total movement of grain, 1865–84	6
2(a)	The relationship of grain exports to net harvest, 1860–79	7
(b)	The relative importance of domestic and export markets, 1865–84	8
3	Measurements of the potential grain surplus, 1857–79	9
4	Domestic retentions of grain, per head, 1851–95	13
5	Trends in grain production—deficit provinces, 1851–82	17
6	Trends in grain production—surplus provinces, 1851–82	18
7	Annual consumption of grain in Moscow, 1789–1876	27
8	Annual consumption of grain in St Petersburg, 1775–1878	29
9	Net exports of grain from the Volga provinces, 1863–82	38
10	The relative advantages of transport by rail, water and cart in 1872	40
11	Transport of cotton to European Russia via Kazalinsk and Orenburg, 1883–92	57
12	Despatches of cotton from Uzun-Ada and Krasnovodsk and total supply of cotton to European Russia from Central Asia, 1887–1915	58
13	Calculation of cotton transport on the Tashkent Railway, 1906–14	60
14	Sources of cotton traffic on Central Asian Railway via Krasnovodsk, 1911–14	61

15	Export and import of grain to and from Uzun-Ada, 1888–93	64
16	Grain despatches on Central Asian Railway and arrivals in Fergana valley, 1899–1912	66
17	Grain traffic on Tashkent Railway, 1905–13	67
18	Transport of grain to Turkestan and Fergana via Tashkent Railway, 1906–13	69
19	Increase of sowings of cotton in Turkestan, 1909–14	71
20	Percentage share of the prime hauliers in total national transport, 1913–67	86
21	Average length of haul by principal media, 1969	87
22	Distribution of rail freight by distance moved	92
23	Share of goods turnover by sea areas in 1960	100
24	U.S.S.R. international airport runways	159
25	Freight transported in the U.S.S.R. and pre-revolutionary Russia by common carriers, 1913–71	166
26	Total freight movement in the U.S.S.R. and pre-revolutionary Russia by common carriers, 1913–71	167
27	Passengers carried in the U.S.S.R. and pre-revolutionary Russia by common carriers, 1913–71	171
28	Total passenger movement in the U.S.S.R. and pre-revolutionary Russia by common carriers, 1913–71	172

Contributors

Terence Armstrong, M.A., Ph.D. (Cantab), Hon. LL.D. (McGill), is Assistant Director of Research, Scott Polar Research Institute, University of Cambridge

Richard Greenwood, M.A. (Cantab), F.R.G.S., is Professor of Geography, University College, Swansea

Roy E. H. Mellor, B.A. (Manchester), is Professor of Geography, University of Aberdeen

Derek W. Spring, B.A., Ph.D. (London), is Lecturer in Russian and East European Economic History, University of Nottingham

Leslie Symons, B.Sc.(Econ.), (London), Ph.D. (Belfast), is Reader in the Geography of Russia, University College, Swansea

Colin White, B.A.(Cantab), is Lecturer in Russian Economic History, University College, Swansea

Note on Transliteration and Rendering of Names

The system of transliteration used is that recommended by the U.S. Board on Geographical Names and is used throughout, not merely for place names. Some difficulties occur, however, in quoting translations. Thus, authors and titles of works that are available in translation are in some cases quoted in the system adopted by the translator in order to facilitate reference. Where reference is made to the original Russian source, however, transliteration follows the system here adopted.

For simplicity, soft and hard signs have been omitted in the text except where a Russian term is given in italics. They are used in names and titles in the bibliographies. Normally, use of Russian terms follows English usage, for example in the plural form, 'oblasts' rather than *'oblasti'*. Place names are given in direct transliteration from Russian, except for the omission of soft and hard signs and special cases such as Moscow and Georgia, and regional names such as Transcaucasia, where English forms are in general use. Similarly a few other words are regarded as anglicised for spelling purposes, for example, *'pood'* is preferred to *'pud'*.

Abbreviations follow English or Russian style according to whether they signify translated or transliterated forms, for example, U.S.S.R., SSSR.

Weights and Measures

The metric system is used in this book except for figures in the historical section where old Russian measures are used. Following Soviet practice the term 'milliard' is used for one thousand millions.

1 metre = 1·0936 yards
1 kilometre = 0·6214 miles
1 square metre = 10·764 square feet
1 hectare = 2·47 acres
1 square kilometre = 247·1 acres (0·386 square miles)
1 kilogram = 2·205 pounds
1 centner = 100 kilograms or 220·5 pounds
1 metric ton or tonne = 1000 kilograms or 2204·6 pounds
1 chetvert = about 6 bushels
1 desiatin (or desyatina) = 2·7 acres
1 pud or pood = 16·38 kilograms or 36·1 pounds
1 verst = 1·067 kilometres or 0·66 miles

Introduction

There can be no denying the fundamental importance of transport improvement to the economic development of tsarist Russia and the Soviet Union. Reference to the sheer size of the territory concerned immediately predisposes an observer to this view. With its present borders the U.S.S.R. is about ninety times the size of the United Kingdom and three times the area of the next largest country in the world. Two other problems complement the simple fact of distance.

Firstly, a number of important disadvantages characterise the natural transport infrastructure. Secondly, the demands imposed on this system were magnified by the dispersion of resources, population and export outlets. Without this dispersion the required input of transport services would have been relatively much less significant. In a country of self-sufficient communities transport needs are small. The demand for the services of a transport system is, therefore, closely linked with the necessity and the desirability of regional specialisation.

However, we must first consider the failings of the natural transport infrastructure. This consists of waterways and overland routes along which either the forces of nature, wind and current, or human and animal traction provide the motive power. Some of the failings referred to are topographical. The ratio of coastline to inland frontier is surprisingly low and from the viewpoint of transportation the effective ratio is much lower due to ice in winter. Most Russian rivers flow in a north–south direction which contrasts with the east–west orientation of the country and some, because of their width and direction of flow, offer important obstacles to overland movement. Moreover, most of these rivers flow either into land-locked seas, such as the Caspian or Aral seas, or into the Arctic. Although they are long, many flow sluggishly or are interrupted by frequent rapids and overflow their banks in spring. Naturally, some areas are better endowed than others. The waterway 'from the Varangians to the Greeks', along the Dnieper, Lovat, Volkhov and Neva rivers, was the basis for the flourishing Kievan confederation of the ninth, tenth and eleventh centuries. Muscovy was conveniently situated near the sources of a number of important rivers, including the Volga, Don, Dnieper, Oka and West Dvina.

Climatically the situation throughout the Russian lands is extreme. Most waterways are frozen over for a considerable part of the year. Even the Black Sea ports fall within the boundary of ice in the winter months. On the other hand, the spring thaw reduces dirt roads to a quagmire. Overland movement before the nineteenth century was easiest in winter when snow cover afforded an ideal surface for sledges. However, there existed the considerable risk of becoming hopelessly lost in a blizzard. Furthermore, large areas in the south and east were bereft of the materials necessary for constructing metalled or even corduroyed highways (see page 98). Clearly the rigours and vagaries of climate meant that any commercial transaction involving transport of commodities over a significant distance also involved a high degree of risk.

These then are some of the difficulties inherent in the supply of transport services. On the credit side there were a large number of unemployed peasant cultivators, especially in the winter months, who provided the cheap human traction necessary to drag the barges along the rivers or cart the luxury goods and estate products between the town and country residences of the nobility, a relatively cheap way of overcoming the problem of distance if one ignores the cost in human terms. Oxen and horses provided more efficient motive power but were often more expensive than human beings.

What of the demand for these transport services? Most Russian households, with the exception of the richer landlords, lived at the subsistence level even in the nineteenth century. In this condition households needed to buy little on the market.

Salt, timber and some metal products, usually very few, can be considered necessary purchases, having to be transported from areas where they were available. In Russia the distances involved in such transport were often very great.

Climatic and soil conditions divide the area of European Russia into a number of agricultural belts running from east to west. Historically the fundamental distinction lay between the treeless steppe of the south and south-east with its rich black-earth and the forests or marshes of the north with their leached, acid podzols. Thus the southern zones had to draw on the northern forests for timber and charcoal, which would be shipped by the Volga and its tributaries or other rivers connecting the different zones. The most populated region around Moscow lay in an intermediate zone where the agricultural potential was weak, with poor soils and a short growing season which limited the range of crops. The pattern of seasonal work made labour costly, with a long winter during which livestock had to be fed. Fortunately, as Russian control pushed further and further south in the course of the seventeenth and eighteenth centuries, land of higher productivity was brought into cultivation and increased supplies of grain and other crops became available. As a result there appeared a marked regional specialisation between the central 'industrial' area with a grain deficit and the grain surplus area of the *chernozem* (blackearth) belt. A considerable trade in grain developed between these two areas.

On the other hand, the development of the cotton textile industry in the central industrial area, mainly around Ivanovo, also brought with it the import first of British cotton yarn and then of American raw cotton. The establishment of Russian control over the Central Asian khanates in the second half of the nineteenth century made it possible to reduce this dependence upon imports. The irrigated land of Bukhara and the Fergana valley was eminently suitable for cotton culture. However, grain was still needed for the subsistence of the cotton producers. Therefore we can see the potential for a triangular trade between the central industrial regions, Central Asia and the south and south-eastern steppes.

So far we have referred to agricultural products for which there is a reasonable freedom of choice in location. For mineral products this choice is very limited. Probably the two key products in the industrial transformation of an economy are iron and coal. Traditionally the Ural mountains had been the source of Russia's iron ore but this had been smelted with local supplies of charcoal. The Urals were remote both from centres of consumption for metal products and from potential sources of coal. These two factors were largely responsible for the relative decline of the Russian iron industry in the second half of the eighteenth and the first half of the nineteenth centuries. There were alternative deposits of iron ore at Krivoy Rog and of coal in the Donbas, but even they were separated by 500 kilometres. Moreover, they were also relatively remote from Moscow and the main centres of population. The rivers in the area were of only limited value in this connection. Therefore the establishment of the Ukrainian metallurgical base was conditional upon the construction of rail links between this area and Moscow, achieved in the 1860s and 1870s, and on the connection of the two raw-material sources, effected by the Yekaterinin railway in the mid-1880s.

The establishment of a second metallurgical base in the Kuzbas in the first Five-Year Plan was itself dependent upon the prior construction of the rail link between the Ural iron ore and the Kuzbas coking coal.

Without these rail links neither of these developments would have been feasible. As a result of the first, coal quite quickly replaced grain as the main item transported on the railways. For products with such a low value-to-bulk ratio, transport improvement was clearly crucial. The export of grain and the movement of coal to the iron-ore deposits were essential prerequisites of the economic development which occurred in Russia before the Revolution of 1917.

In the Soviet period the discovery of large new deposits of valuable raw materials in remote areas, such as eastern Siberia, has brought a new regional problem to the fore. Should the Soviet authorities transfer the resources to the populated areas, or people to the resources? Once again regional policy, or regional specialisation, has serious implications for transport policy. In a planned economy transport improvement is realised in a broader context of regional policy. In a market economy regional specialisation is to some extent the unexpected result of transport improvement. The first two chapters of this book illustrate the latter point, the third the former.

The broad conclusion of the previous discussion is that a country can only be judged as rich in resources in the context of the accessibility of these resources. By this criterion Russia was not resource-rich. Improved transportation was crucial to its economic development.

The history of transport improvement starts in Russia, as in the industrial countries of western Europe, with canal construction, unless one considers the attainment of control over natural transport routes as an aspect of such improvement. Peter the Great instituted a programme of canal construction which proved to be abortive, with the exception of the famous Vishnevolotskaya system linking the Volga with St Petersburg. Canal construction was resumed at the end of the eighteenth century and continued into the nineteenth century when again the main objective was the improvement of communications between the Volga and St Petersburg. However, most of the main waterways of European Russia were linked by canals at this time.

Interest in the railways grew apace in the 1830s and the decision to build the first line, the short Tsarskoye Selo line, was taken, mainly at the prompting of Nicholas I, in 1835. The first economically important line was the Moscow–St Petersburg railway, later called the Nikolayev railway, constructed between 1842 and 1851. As with many other reforms, or at least changes in policy, a strong conviction in the necessity of much more widespread railway construction was a product of defeat in the Crimean War. The major problem was finance. This necessitated considerable government participation in railway construction and operation, although the degree of participation varied from decade to decade. The motivation in railway construction was in many cases probably as much military as economic. However, many of the lines constructed obviously had great economic significance, and the first two chapters in this book approach the issue from this point of view. The focus of interest is the economic significance of the lines, not the primary motivation behind their construction.

The pace of railway construction slackened off during the First World War and in the period of disorganisation following it. Moreover, although the Soviet authorities attached great importance to the railways they gave investment in railway construction a low priority and required of the existing railways the fullest possible utilisation. To some extent the railway network inherited from the tsarist period was not well adapted to Soviet demands. It had an export orientation which did not accord well with the autarkic goals of the early Five-Year Plans. Soviet railway development has been orientated towards economic development, particularly in the eastern regions, and the U.S.S.R. is outstanding among economically advanced countries in the scale of new railway construction at the present time.

The later revolution in transport services associated with motor vehicles has only belatedly been accepted by the Soviet government, and the U.S.S.R. is still far in arrears of other developed countries in the quality and quantity of roads, vehicles and general infrastructure for road transport. On the other hand, the Soviet effort in development of air transport has compared favourably with world trends, while much attention has also been given to shipping in recent years.

Interest in the continuing significance of transport improvement for the economic development of both the tsarist empire and the Soviet Union has been limited and unsustained. The resulting literature, both in English and Russian, has been poor. Inside Russia a number of interesting studies were published during the tsarist period, but since then very little of any note has appeared. This is to be expected as the role of transport improvement is considered in its broader Marxist context of the extension of the market. Secondary importance is attached to transport improvement *per se*. Furthermore, western writers have followed their Soviet counterparts in giving far more weight to institutional changes such as the Emancipation of the serfs, or Stolypin's reforms, in explaining Russia's growth performance, than to railway construction or transport improvement in general. A strong assertion of the importance of transport improvement made by Baykov in an article published in 1954 has been largely ignored. Recent work has tended to concentrate on the debates which preceded the construction of the first railways rather than on their economic significance. However, Holland Hunter's work on the Soviet period has certainly cast greater light on more recent developments.

The paucity of material on Russian transportation is rather surprising to the economic historian in view of the recent work done on the role of railway construction in the economic development of the U.S.A. In his theory of 'take-off' Rostow offered the railways as the leading sector in both the U.S.A. and Russia. For a long period general statements have been made about the importance of forward, backward and lateral linkages which accompanied railway construction. Forward linkage effects are those experienced by the consumer of transport services. These effects are translated into a change in the cost or quality of the services provided. Cost changes occur together with changes in the speed, regularity and risk of movement. Backward linkage effects are experienced by the suppliers of materials or goods necessary for railway construction or operation. Clearly the iron industry was likely to be that most affected. Lateral linkage effects arose from the considerable investment effort involved, which not only might attract otherwise unavailable investible funds from abroad but may also, through multiplier-accelerator effects, stimulate the growth of the economy. Fogel and Fishlow for the U.S.A. and Hawke for England have tried to quantify these effects.

In quantifying the forward linkage effects Fogel introduced the concept of 'social saving'. This measured the extra cost of using the next best alternative transport system in the absence of the railways, clearly a combination of water and overland transport. Fogel's conclusion was that the total social saving in 1890 was equal to only one year's growth. Moreover, the backward linkage effects were much weaker than usually implied. The amount of iron required for rails, locomotives and rolling stock was less than that demanded for nail production.

These broad conclusions were confirmed by the work of Fishlow and by Hawke's work on England. Quite clearly the methods and conclusions have important implications for the Russian case. With respect to Russia two different attitudes to Fogel's work are valid. One could argue that Fogel neglects some very important benefits of transport improvement, the so-called externalities, which are particularly significant in the Russian case. For example, the first paper argues that the construction of the railways released a grain surplus, already in existence, for export. Alternatively one might argue that application of a similar method would yield very different results when applied to Russia. Our brief survey of the climatic and topographical conditions in Russia has indicated that transport cost was likely to be a much more important component of total costs than elsewhere. The high potential for regional specialisation also suggests the significance of transport services for economic development. For these reasons we might expect a higher social saving than for the U.S.A. and England. It is certainly surprising that Fogel's work has not awakened greater interest in the construction and operation of the Russian railways.

The method, an application of cost-benefit analysis to the past, could, of course, be used to assess the significance of any transport innovation, such as the introduction of the internal combustion engine. However, the higher level of economic development at the time of this innovation would lead one to expect that its contribution to economic growth would be relatively less important than that of railway construction.

The papers in this volume are not intended to give a comprehensive account of the development of all transport systems in the Russian Empire and the Soviet Union, but rather to examine some of their salient features. The railways are given a place commensurate with their importance. The first two chapters deal with different aspects of railway development before the Revolution. Chapter 3 describes the role of transport in the broader planning system and dwells in detail on the policy of the Soviet authorities towards the railways. Developments in the merchant marine, the use of the northern sea route and in aviation are recounted in Chapters 4, 5 and 6 respectively. The major lacuna is road transport, although this neglect corresponds with the relative priority accorded by the Soviet authorities in the past to this mode of transport. Its growing importance is placed in perspective to the other forms of transport in the concluding chapter, which also includes the statistics for the period between the beginning of the Soviet Five-Year Plans and the present. Also in this final chapter will be found the latest available statistics, published after the other chapters of the book were completed.

If we are to claim that there is a unifying theme for this volume apart from the relationship between economic growth and transport improvement, it might be regional specialisation and the regional orientation of the developed transport systems. Three papers, those on grain production, cotton production and the northern sea route, deal directly with this issue. The other papers treat it indirectly. There are major gaps in the treatment. For example, there is no discussion of the relationship between transport improvement and those industries supplying transport media with the materials they require for construction or operation. Such gaps, it is hoped, will be filled in later works. Thus aware of its shortcomings, we nevertheless hope that this collection of papers will be found to make a positive contribution to a neglected field.

BIBLIOGRAPHY

Baykov, A., The economic development of Russia, *Economic History Review*, 2nd series, VII, 1954

Fishlow, A., *American railroads and the transformation of the ante-bellum economy*, Cambridge, Mass., 1965

Fogel, R. W., *Railroads and American economic growth: essays in econometric history*, Baltimore, 1964

Hawke, G. R., *Railways and economic growth in England and Wales 1840–1870*, Oxford, 1970

Hunter, H., *Soviet transportation policy*, Cambridge, Mass., 1957

Hunter, H., *Soviet transport experience: its lessons for other countries*, Washington, D.C., 1968

Rostow, W. W., *The stages of economic growth*, Cambridge, Mass., 1963

1

The Impact of Russian Railway Construction on the Market for Grain in the 1860s and 1870s

Soviet economic historians (following Lenin) have viewed transport improvement as a necessary prerequisite of market expansion.[1] In his main work on Russian economic history, *The Development of Capitalism in Russia*, Lenin places railway construction at the head of a number of factors promoting commodity circulation.[2] However, it is clear that the impact of railway construction on commodity circulation is dependent upon its effect on the social division of labour. 'The "market" arises where, and to the extent that, social division of labour and commodity production appear. The dimensions of the market are inseparably connected with the degree of specialisation of social labour.'[3] The role of the railways in increasing commodity circulation and commodity production was reinforced in Russia's case by geographical factors. Given the importance of the agricultural sector in the period under study, soil and climatic conditions still tied the social division of labour to specific regional patterns. This is most obvious

---

[1] See Lyashchenko, 1948, vol. II, 120 and following, Khromov, 207 and following, Lyakhovskiy, 34.
[2] Lenin, vol. 3, 551.
[3] Ibid., vol. 1, 100.

for the main export and consumption good, grain. During the 1860s and 1870s grain was by far the most important single commodity transported on the railways and for this reason alone merits special attention.[1]

In this paper there is, firstly, an attempt to discover whether the proportion of grain marketed in the 1860s and 1870s was rising. Secondly, the proposition that transport improvement leads to a greater concentration of grain production on high-yielding provinces, that is, increased regional specialisation, and thereby to an increase in the proportion of the harvest marketed, is tested. Thirdly, changes in the size and supply pattern of the main domestic grain markets, Moscow and St Petersburg, are analysed.

The periodisation adopted is justified by the fact that the first important wave of railway construction, 1868–74,[2] established a network of railways supplying Moscow and St Petersburg with grain and the beginnings of a network of export outlets. These lines served mainly the central agricultural region which benefited most from them. It was only in the 1860s that grain movements on the railways became at all significant. Furthermore, the market for grain was buoyant until about 1880 when the opening of the 'new' countries by railway and steamship began to lower grain prices within Russia.[3] Conditions during the 1860s and 1870s were in general favourable to the expansion of the grain market. After 1880 the

---

[1] In 1874 grain accounted for 44·8 per cent of total pood-versts of freight transportation. Coal and timber, the next two most important items, each accounted for just over 5 per cent. Bliokh, vol. I, Table XXIII.

[2] The railway network grew annually by the following distances, in versts:

1838	25	1861	463	1871	2626
1845	20	1862	1117	1872	510
1847	23	1863	197	1873	1974
1848	288	1864	90	1874	1741
1850	111	1865	208	1875	712
1851	470	1866	676	1876	512
1853	42	1867	447	1877	1055
1857	113	1868	1174	1878	1179
1859	159	1869	1190	1879	628
1860	240	1870	2441	1880	125

Sources: Bliokh, vol. I, 62–65; Stat. sborn. for relevant years.

[3] See Yegiazarova, Chapter 3.

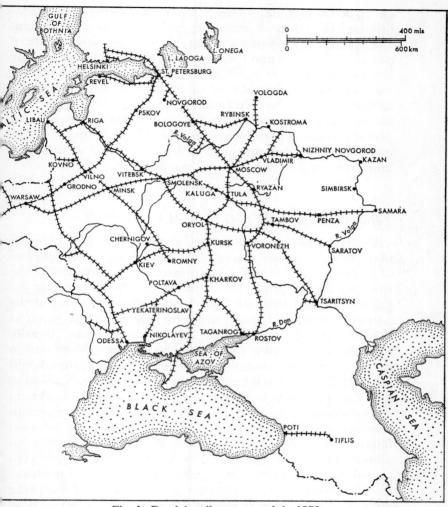

Fig. 1. Russia's railway network in 1875

situation changed dramatically. This change is illustrated by the failure of the important Ryazan–Kozlov railway to sustain its movements of grain in the 1880s.[1]

One complicating factor must be mentioned. The Emancipation of the serfs in 1861 makes it difficult to disentangle the

---

[1] See Lyakhovskiy, 50.

effects of transport improvement and Emancipation. Emancipation had the effect of increasing grain movements in general and on the railways in particular. Firstly, the monetisation of peasant obligations and their increased financial burdens made it necessary for them to sell more grain. Secondly, free carting was no longer available to the landlords, which changed relative transport costs in favour of the railways. Fortunately, all these influences worked in the same direction. During this period increasing grain prices and the effects of Emancipation established an environment in which the impact of transport improvement should have been maximised.

THE RAILWAYS AND GRAIN MARKETINGS

Firstly, what was happening to the proportion of grain marketed? There are two possible approaches to the problem of defining the size of the grain market. One can define the consumption needs of the producers and by subtracting the grain consumed from total production derive the size of the potential surplus.[1] Ideally, the household should be taken as the relevant unit. However, this is not feasible. Usually surpluses and deficits are calculated by province. Moreover, to derive an accurate estimation of the potential surplus one should subtract the grain required for the following year's sowings, and that necessary to maintain man and beast at minimum consumption levels. Usually, however, the average consumption per head for European Russia is taken as the relevant figure. Since it is argued that before improved communication allowed more of the surplus to be marketed the average consumed exceeded minimum requirements, there is a tendency to underestimate the potential surplus. Furthermore, the amount of grain used in distilling is also subtracted from total production. In the studies quoted, therefore, the potential surplus is underestimated by the size of intra-provincial marketings, by the amount of grain consumed over and above the necessary minimum and by the quantity of grain consumed

---

[1] This approach is adopted by Vilson, Bliokh and the writers of the *Voyenno-statisticheskiy sbornik* of 1871.

in distilling. However, since the railways should have their biggest impact on the long-distance movement of grain rather than intra-provincial movements, our analysis should at least isolate the area of greatest change.

Alternatively one can start with movement of grain by rail, water or cart and use this as a measure of total marketings.[1] Again the relevant data indicate only inter-provincial movements.

Potential surplus and marketings differ by the size of changes in local stocks and the amount by which local consumption exceeds the necessary minimum. This difference is likely to fluctuate from year to year if consumption norms vary with the size of the net harvest.

Both methods are subject to considerable error. Total production figures are suspect; consumption norms vary according to the authority.[2] There is inadequate coverage of grain movements by rail or water, and figures for movement by cart are only tentative estimates. However, on the basis of a number of contemporary studies we can attempt to indicate the extent of marketings or at least the underlying trend. In

---

[1] Studies by Bliokh, Shmeyn and Borkovskiy exemplify this approach.

[2] Opinions on average consumption per head differ greatly. Sometimes these differences can be explained by regional disparities or by a differing coverage. For example, Bliokh takes 1·4 chetverts per head for human consumption and, according to the area, 4·2 and 1·2 chetverts for horses. He takes separate account of distilling. Vilson takes 2·2 chetverts per person and 2·5 chetverts per horse. Distilling is included in the first. Chaslavskiy agreed with the latter figure but raises human consumption to 2·6 chetverts per head. Presumably this includes distilling, but it applies to Kursk province in the heart of the agricultural region. The authors of the *Voyenno-statisticheskiy sbornik* take 2·15 chetverts per person and allow separately for distilling, but not for animal consumption. Chuprov and Borkovskiy approach the problem from a different angle. Chuprov subtracts seed grain and exports from the total harvest and divides the remainder by the figure for population. The result is a level of total domestic consumption of about 2·5 chetverts per head. Borkovskiy looks at grain available for consumption in the basin of the Upper Volga and divides by population to reach a figure just under 2 chetverts. Both these figures cover all demands.

It does appear that average consumption per head, including distilling, is about 2 chetverts. Inclusion of consumption by horses raises the figure to about 2·5 chetverts per head.

Sources: Bliokh, vol. II, 21–24; Chaslavskiy, Part II, 233; *Voyenno-stat. sborn.*, 248; Chuprov, vol. I, 248, footnote; Borkovskiy, *Izsled* 39.

Table 1 there is an attempt to estimate total grain movements. In Tables 2(a) and 2(b) these movements are compared with the net harvest and divided into movement for export and for domestic consumption.

TABLE 1

TOTAL MOVEMENT OF GRAIN

(in 50 provinces of European Russia—Poland, Finland and the Caucasus excluded) (in millions of poods)

Year	Mode of transport			
	1 Water	2 Rail	3 Cart	4 Total
1865–7 average[1]	113	36	87	237
1874[2]	124	196	60	380
1875[3]	119	215	60	394
1876[4]	129(120)	249(233)	60	411
1878[5]	(96)	(391)	60	539
1880[5]	(98)	(292)	60	426
1874, 5, 6, 8 and 1880 average				430
1880, 1882, 1884 average[6]	120(110)	(354)	60	509

In column 1 the figure for 1874 is the average for the two following years, 1875 and 1876. Bliokh's original figures cover movement on only a few main waterways. The figures in brackets represent movement of the six main items only—wheat, rye, barley, oats, wheat flour and rye flour. In this article grain is understood as comprising all grains and flours.

In column 2 the figures in brackets represent movement of the eight main items—those covered in the six above plus maize and buckwheat.

Column 3 poses some problems. The figures for 1874, 5, 6, 8 and 1880 are based on the following assumptions. According to Vilson,[7] in 1865 grain movements by cart were distributed in the following way; in the Kama region 7 million poods, in the central region 15 million, in the west 10 million and in the south 55 million. The main impact of railway construction was to reduce movements by cart to Moscow and the west to a mere trickle, mainly supplying parts of northern Ryazan and Vladimir provinces.[8] We assume a drop from 25 million to 5 million. The Kama region was unaffected by railway construction and therefore we assume an increase from 7 million to 10 million. Estimation of movement in the south is much more difficult. There is reference to a reduction of 13 million poods in grain carted to Odessa after a railway had been completed linking Odessa with its hinterland.[9] However, in 1874 exports of wheat alone exceeded arrivals by rail by as much as 23·9 million poods. Some of this arrived by water but allowances for the other items and for consumption in the ports would probably double the figure. According to Pokrovskiy,[10] in 1895, when there were far more railways in the area, 115 million poods were carted to ports on the Black Sea and Sea of Azov for export. This was slightly less than one-third of the total exported. In the 1870s, 90 million poods were exported through these ports. Therefore, it seems that at least 30 million poods would have been moved by land for export. Furthermore, the total urban population of New Russia was probably about the same as Moscow's, where about 20 million poods of grain were consumed.[11] Conservatively, we allow 30 million poods moved for export and 15 million for urban consumption, a total of 45 million poods moved in the south. This leaves us with a grand total of 60 million poods. Fedorov's estimate of 60 million poods on average for the years 1880, 1882 and 1884 includes only movements of grain to the ports and to overland border points.

In column 4 transhipments between the different modes of transport are netted out. Allowance is made for incomplete coverage in 1876, 1878 and 1880.

Sources: [1] Vilson, 171.
[2] Bliokh, vol. II, 87 and 88.
[3] Shmeyn, V and VI.
[4] *Stat. sborn.*, vyp. 3, 16 and 17.
[5] Bilimovich, statistical appendix, Tables 3, 9 and 10.
[6] Fedorov, 7 and 8.
[7] Vilson, 171.
[8] Chaslavskiy, Part I, 177 and 178.
[9] *Voyenno-stat. sborn.*, 582.
[10] Pokrovskiy, 8 and 9.
[11] See Table 7.

TABLE 2(a)

THE RELATIONSHIP OF GRAIN EXPORTS TO NET HARVEST
(in millions of chetverts)

1	2	3	4	5
Years	Net harvest[1]	Years	Exports[2]	Exports/Net harvest per cent
1860–4	152	1861–5	9.3	6.1
1865–9	139	1866–70	14.9	10.7
1870–4	186	1871–5	22.5	12.1
1875–9	179	1876–80	34.0	19.0
1860–9	145	1861–70	12.1	8.3
1870–9	183	1871–80	28.3	15.4

Sources: [1] Nifontov, 45.
Pokrovskiy, 105.

TABLE 2(b)

THE RELATIVE IMPORTANCE OF DOMESTIC AND EXPORT MARKETS

(in millions of chetverts)

1	2	3	4	5	6	7	8
					As percentage of column 2		
Years	Net harvest	Total movements[1]	Exports	Domestic movements	Col. 3	Col. 4	Col. 5
1865–7	133	26·3	13·3	13·0	19·8	10·0	9·8
1874–5–6	178	47·9	29·3	18·6	26·9	16·5	10·4
1878 and 1880 1880, 1882 and 1884	189	56·6	33·3	23·3	29·9	17·6	12·3

The figures for the net harvest are generally considered to be underestimated by a fairly large amount. Moreover, they are certainly unreliable. The first factor helps us in that the degree of underestimation almost certainly declined over time. Therefore, the figures for marketability are more exaggerated for the earlier periods. In looking for the direction and degree of change in marketings we almost certainly err on the conservative side. However, the unreliability of the figures must make us chary of drawing any conclusions from the absolute size of the figures. Exports of a given year are related to the harvest of the previous year. The same applies to Table 3, where the harvest figures in column 2 are for the years preceding those stated.

Conversion from chetverts to poods or vice-versa is done at the rate of one chetvert equalling nine poods, irrespective of the item involved. This introduces an element of inaccuracy.

A number of conclusions can be drawn from these tables:
1. The increase in the net harvest more than kept pace with population increase. The annual rate of increase during the 1860s and 1870s was 1·5 per cent and 1·3 per cent respectively.[2] However, after allowance is made for increased exports it seems that consumption levels must have remained constant.
2. The total movement of grain increased significantly both in absolute volume and as a proportion of the net harvest.
3. There was a striking increase in grain exports, which

---

[1] See Table 1.

[2] Goldsmith's figures apply to the period 1860–83. Goldsmith, 449 and 493.

doubled as a proportion of the net harvest between the 1860s and 1870s.

4. There appears to have been some increase in the share of the harvest directed to meet the domestic needs of deficit provinces. The argument has been put in such a way that the increase calculated is a minimum.

If we consider the two estimates of the size of the potential surplus, by Vilson and Bliokh, we can draw some tentative conclusions about the nature of the impact of transport improvement on marketings.

TABLE 3

MEASUREMENTS OF THE POTENTIAL GRAIN SURPLUS
(in millions of chetverts)

1	2	3	4	5	6	7	8
Years	Total Surplus	Domestic Deficit	Exports	Residual	Net harvest	Col. 2 as percentage of Col. 6	Col. 5 as percentage of Col. 6
1857–63[1]	42·9	7·9	9·4	25·6	151	28	17
1870–2[2]	50·2	14·1	20·4	15·7	184	27	8
1875–9	50·2	14·1	33·8	2·3	179	28	1

The total surplus and domestic deficit are the aggregation of surpluses and deficits by province. The last line allows only for the increase in export of grain.

Sources: [1] Vilson, 109 and 110.
         [2] Bliokh, vol. II, 25.

Almost certainly Vilson's calculation of the domestic deficit is an underestimate, inconsistent with his own figures for grain movements, 1865–7. Substitution of a more reasonable figure of about 13 million chetverts would reduce the ratio, residual to harvest, to 14 per cent. However, if we allowed for underestimation of the harvest the residual would be higher. It is possible that the figures are so inaccurate that no conclusions can be drawn, but the stability of the potential surplus is an argument against this view.

The trend interesting us here is the reduction in the ratio of residual supply of grain to net harvest. It has been argued that violent fluctuations of local harvests and poor communications compartmentalised local grain markets and made necessary the accumulation of substantial stocks of grain to meet deficits in years of poor harvest.[1] Much grain must have been spoiled in store or transit, or even stolen. Consumption in rural areas might have been considerably higher than usually assumed, particularly in good years. The argument advanced here is that the diminished importance of the residual was the result of improved communications. The need to hold stocks against bad times was reduced. Less grain was lost in store or transit, or stolen. There may also have been a tendency for consumption to be equalised between rural and urban areas. Is there any evidence for this argument?

We can start by showing that there still existed an unmarketed surplus at the end of the 1870s and then try to show that it had declined over the preceding years.

In theory, marketings and potential surplus differ by the amount of grain put into local stocks or simply lost, spoiled and stolen before transport. Since the average level of consumption is assumed to be the same for all provinces, the difference is not so easily explained. As one might anticipate, the evidence suggests that consumption of grain per head was higher in producing provinces than in consuming provinces.[2] Application of an average figure to all provinces would tend to exaggerate the size of both actual surpluses and actual deficits. Such unevenness of consumption levels, however, was itself largely the result of difficulties in communication. A fear of

---

[1] Zablotskiy, for example, argued that the grain surplus of 12·5 million poods estimated by Protopopov for an average year was only sufficient to meet a month's domestic demand. See Zablotskiy, 38 and 39. For a discussion of the argument that this surplus was required to replenish stocks for years of poor harvest, see Blum, 333.

[2] Borkovskiy's figure for the Upper Volga basin and Chaslavskiy's for Kursk province would tend to confirm this view. Borkovskiy has also calculated net retentions by region on the basis of a reduction of all grains to poods of rye according to their nutritional content. The average is 15·37 poods. The resulting regional breakdown is as follows:

Polish provinces	10·1	North-east	16·2
North-west	10·4	Central blackearth	16·7
South-west	12·7	South	16·8
North	12·7	South-east	(?) 22·4
Central non-blackearth	15·4		

Source: Borkovskiy, *Ocherk*. . . , 11.

local famine would operate, not only to encourage the maintenance of considerable stocks of grain, but also to keep production above what was strictly necessary to meet the needs of an average year (climatically speaking). This would almost certainly be reflected in higher consumption per head. It seems reasonable, therefore, to regard the difference between potential surplus and marketings as evidence of the existence of an additional potentially marketable surplus. We would expect the replenishment of stocks, the incurring of losses and the maintenance of above-average consumption to be reflected in a short-fall of net exports from producing provinces below the potential surplus. Comparison of Bliokh's figures for the potential surplus with Borkovskiy's calculation of net exports confirms this expectation (see Tables 5 and 6). However, the period covered by Bliokh's analysis was one of above-average harvests, particularly in the central blackearth belt.[1] The average harvest 1870–2 was about the same as that in 1878, 1800 and 1882. Since, in the meantime, population had risen, one would expect the potential surplus to have fallen somewhat. However, the analysis so far seems to indicate that some at least of the potential surplus was still unmarketed at the end of the 1870s.

Similarly, for consuming provinces the comparison suggests that, for most, Bliokh exaggerated the actual deficits. In these provinces consumption levels were probably lower than average. In this area the two major exceptions were Moscow and St Petersburg provinces where net imports exceeded the potential deficit by large amounts. Three factors could explain why net imports of grain were so much higher at the end of the 1870s than the potential deficit at the beginning of the decade: firstly, the rapid increase in the degree of urbanisation of these provinces during the 1870s;[2] secondly, a continuing

---

[1] See Nifontov, 56.
[2] The following facts support our contention:
   (*i*) The period, 1867–85, was marked by the fastest growth of urban population of the whole Tsarist era.
   (*ii*) A tendency for the share of urban population represented by Moscow and St Petersburg to decline was reversed during this period.
   (*iii*) Both cities began a period of particularly rapid growth in population; Moscow probably in the 1860s and St Petersburg in the 1870s. The

contraction of the sown area; thirdly, an increase in consumption levels made possible by the railway network, which was at its densest in these two provinces. Clearly, railway development had a major impact on these two provinces, if only in allowing a more rapid rate of urbanisation.

Unhappily, there is no possibility of using a similar analysis to show changes in the utilisation of the potential surplus over time. We have merely shown the existence of some unutilised potential surplus at the end of the 1870s.

Some direct evidence exists in comments made by contemporaries on various factors mentioned in the argument.[1] However, there is a strong line of indirect argument.

Our previous analysis has suggested that the main area of expansion in sale of grain was abroad. Exports of grain rose impressively as a proportion of the harvest, from 5 per cent of the gross harvest at the beginning of the 1860s to over 14 per cent between 1876–9 and from 6 per cent of the net harvest, 1860–4, to 18·9 per cent in 1875–9.[2] Allowance for an increase in the proportion of grain exported all but accounts for the reduction in the size of the ratio, residual to harvest, in Table 3. In fact, a residual of 14 per cent would be just about sufficient to meet the increased export demand. After allowing for exports of about 34 million poods in the period 1875–9, we are left with a mere 2 million poods to accommodate an increase in the domestic deficit. Consequently, there is very little room for any increase in total imports by deficit provinces during the late 1860s and 1870s.

In this situation increased marketing of a potential surplus should be reflected in a decrease in net domestic retentions of grain per head, provided productivity levels have not risen sufficiently to compensate for increased exports. However, these retentions, which certainly appear to show a marked

---

population of both provinces grew at a rate well above the average.
Sources: (i) Harris, 234–5,
(ii) Rashin, 111,
(iii) Rashin, 44, 45 and 113 and following.

[1] On accumulation of stocks in Kursk province, see Chaslavskiy, Part II, 22 and 23. On reduction of average consumption levels after the construction of railways in Kursk, see Chaslavskiy, Part II, 32.

[2] Pokrovskiy, 6.

fall to a new plateau (Table 4), could decline for other reasons than that railway construction had stimulated exports. The greater monetisation of obligations which followed the Emancipation imposed upon the peasants the need to sell grain in the market, notwithstanding a poor harvest or low grain prices. In these cases non-payment of taxes or redemption dues did act as a safety valve and reduce the fall in retentions per head.[1] Nevertheless, it might be argued that the low retention ratios of 1865–9 and 1875–6 to 1879–80 are explained by runs of poor harvests and those after 1880–81 by falling grain prices. However, this argument would not explain why, with a stable level of the harvest, exports rose so dramatically between the first and second halves of the 1870s.

TABLE 4
DOMESTIC RETENTIONS OF GRAIN, PER HEAD
(in chetverts)

1851–4	2·65			1875/6–1879/80	2·01
1855–9	2·11	1870/1–1874/5	2·40	1880/1–1884/5	2·09
1860–4	2·33			1885/6–1889/90	1·96
1865–9	1·98			1890/1–1894/5	1·99
Average for 1851–74/5				Average for 1870/1–1894/5	
2·29				2·09	

Notes: After 1870 the agricultural year is taken as the relevant time period. Allowance has been made for the loss of exports during the Crimean War, affecting the period 1851–4.
Sources: The figures for the period up to 1870/1 are worked out on the basis of Nifontov's figures for net harvest and Rashin's population figures. For those after 1870/1, see Pokrovskiy, 7.

However, if we concentrate our attention on the 1870s we are left with a conviction that improvement in communications released an already existing surplus for export. Retentions of wheat show a very big decline, from 60 per cent in 1870–4 to 44 per cent in 1875–9.[2] In absolute terms retentions show a staggering decline from 136 million poods to 87 million. Despite a fall in the average harvest of about 31 million poods, exports rose by 18 millions on average. For the main consumption grains, rye for human and oats for animal

---

[1] Nifontov, 52.   [2] Pokrovskiy, 15 and following.

consumption, a tendency to increase exports is constrained by the need to meet consumption demands. Nevertheless, the retention ratio declined from 92 per cent and 93 per cent to 87·5 per cent respectively, in absolute terms from 616 to 582 million poods and from 303 to 293 million. However, even for rye, exports rose by 30 million on average whereas the harvest was 4 million poods lower. For oats, exports rose by 19 million poods per annum while the net harvest increased by only 8 million.

Quite clearly something had given a sudden stimulus to exporting. It certainly was not an improvement in the relative profitability of external sales. In the main western markets and in Russia's ports grain prices had already begun their downturn before 1880.[1] Nor could it be argued that there was a sharp deterioration in the financial position of the peasantry, forcing them to sell a higher proportion of their output. Domestic grain prices began to fall after 1880. Only the sudden improvement in communication offered by the construction of the railways could explain this stimulus. It does appear that the railways released an already existing potential surplus. There is little evidence that railway development led to a significant increase in yields per hectare or per labourer, although it did allow new areas of higher yield to be brought under the plough.[2]

THE RAILWAYS AND REGIONAL SPECIALISATION

To what extent did the increase in the proportion of grain marketed result from increased regional specialisation? As transport costs are lowered one would expect to see a contraction of sowings in provinces with a low average yield and an expansion in those with a high average yield.[3] An increase in

---

[1] Yegiazarova, Chapter 3.
[2] See p. 25, footnote 1.
[3] Strictly speaking the relevant factor is the marginal revenue yield. In a situation of perfect competition, perfect factor mobility and no transport costs, the marginal yield of cultivated land would be equalised in all provinces. Where transport costs are greater than zero the price of grain would diverge between surplus and deficit provinces. In deficit provinces grain

Fig. 2. The provinces of tsarist Russia

production would expand, in surplus provinces contract. Therefore, in a situation of diminishing marginal yield the marginal yield in the 'producing' province would be higher than in the 'consuming' province. On the assumption that the pattern of diminishing yields is broadly similar in all provinces, average yield can be taken as a substitute for marginal yield. 'Own' yield does not, however, include the effect of differences in the density of sowings.

the potential deficit for low average yield provinces and an increase in the potential surplus for high average yield provinces would tend to confirm this argument. The greater the density of the rail network constructed in a given province and the larger the reduction in transport costs achieved on existing routes to and from that province, the further one would expect regional specialisation to go. Producing areas by-passed by the railways would also suffer as new supplying areas were linked with their markets. Furthermore, regional specialisation would be accompanied by changes in the whole pattern of supplying existing markets.

Data to test the proposition have been assembled in Tables 5 and 6, an attempt being made to group the provinces by common characteristics as follows:

1. *Metropolitan provinces*
Very high density of railway—a high rate of population increase—a major contraction in grain production—a big increase in the grain deficit, particularly in Moscow province.

2. (*a*) *Core of central industrial region*
High railway density, except in Kaluga—a very low rate of population expansion (Kaluga again an exception)—a major contraction in grain production—a marked increase in the grain deficit.

(*b*) *Periphery of central industrial region*
This group differs from (*a*) by lower railway density, a slightly higher rate of population expansion, a less marked contraction in grain production and only small changes in grain deficits or surpluses.

3. (*a*) *Core provinces of western region*
High railway density (except Pskov province)—fairly rapid rate of population increase—a significant contraction in grain production.

(*b*) *Periphery of* (*a*)
About average railway density—more rapid rate of population increase than (*a*)—some contraction in grain production.

TRENDS IN GRAIN PRODUCTION—DEFICIT PROVINCES

Provinces	Railway density[1]	Population[2]		Change	Grain yield[3]	Cultivated area[4]		Change	Trend in grain sowings[5]	Grain surplus[8]		Net exports[9]	
1	2	3	4	5	6	7	8	9	10	11	12	13	
	Beginning of 1874	1863	1882	(1863–82)	1851–60	1860	1881	(1860–81)	1857/63– 1870/72	1857/63[6]	1870/27	1878, 80 and 82	
	Versts per 1000 sq. versts	Millions		Per cent	Chetverts per chetvert sown	Millions of hectares		Per cent		Millions of chetverts		Millions of chetverts	
Metropolitan Provinces													
Moscow	20.2	1.56	2.14	+37	2.3	1.30	1.07	−18	V.H.−	−1.16	−2.60*	−3.52	
St Petersburg	12.0	1.17	1.62	+39	2.9	0.88	0.69	−21	H.−	−1.57	−1.64*	−3.41	
Central Industrial Provinces													
Vladimir	11.0	1.22	1.35	+11	2.8	2.14	1.77	−17	H.−	−0.22	−0.90*	−0.47	
Tver	8.3	1.52	1.62	+7	2.5	2.07	1.76	−15	V.H.−	−0.25	−1.17*	−0.12	
Yaroslavl	8.4	0.97	1.08	+11	2.7	1.25	0.96	−23	H.−	+0.37	−0.29*	−0.10	
Smolensk	10.1	1.14	1.19	+4	2.3	2.14	1.63	−24	V.H.−	−0.21	−0.83	−0.10	
Kaluga	0	0.96	1.14	+19	2.2	1.66	1.26	−18	V.H.−	+0.12	−0.40*	−0.15	
On Periphery of this area:													
Kostroma	0.9	1.07	1.28	+20	2.8	1.73	1.68	−3	V.H.−	−0.21	−0.40	−0.20	
Nizhniy Novgorod	1.6	1.29	1.43	+11	3.3	1.98	2.21	+11	H.−	+1.29	+0.19	−0.20	
Novgorod	3.1	1.01	1.13	+12	2.6	1.52	1.59	+5	H.−	−0.99	−0.71*	+0.04	
Western Provinces													
Pskov	3.8	0.72	0.89	+24	2.6	1.43	1.21	−15	H.−	−0.11	−0.29*	−0.06	
Mogilev	8.9	0.92	1.15	+25	2.1	1.62	1.35	−17	H.−	−0.27	−0.01*	−0.02	
Vitebsk	12.8	0.78	1.17	+50	2.2	1.30	1.22	−6	V.H.−	−0.50	−0.11*	−0.17	
On Periphery of this area:													
Minsk	5.9	1.00	1.57	+57	2.8	1.93	2.20	+14	L.−	+0.06	−0.28*	+0.21	
Chernigov	8.2	1.49	1.97	+32	2.9	2.84	2.83	—		+0.71	−1.02	+1.07	
Northern Provinces													
Olonets	0	0.30	0.33	+10	3.4	0.31	0.39	+24	L.+	−0.35	−0.29*	−0.11	
Arkhangelsk	0	0.28	0.32	+14	3.3	0.09	0.10	+4	N.A.	−0.53	−0.34*	−0.08	
Vologda	0.2	0.98	1.16	+18	3.4	0.89	0.93	+5	N.A.	−0.99	−0.71	+0.13	
Perm	0	2.14	2.54	+19	3.5	3.22	3.32	+3	N.A.	−0.06	−0.75	+0.42	
Astrakhan	0	0.38	0.45	+18	3.3	0.26	2.13	+719	N.A.	−0.43	−0.78*	−0.12	
Average for all provinces				+27.7	3.2			+30					

TABLE 6

TRENDS IN GRAIN PRODUCTION—SURPLUS PROVINCES

Provinces	Railway density[1] Beginning of 1874 Versts per 100 sq. versts	Population[2]		Change[5] 1863-82 Per cent	Grain yield[3] 1851-60 Chetverts per chetvert sown	Cultivated area[4]		Change 1860-81 Per cent	Trend in grain sowings[5] 1857/63-1870/2	Grain surplus[8]		Net exports[9] 1878, 80 and 82
		1863	1882			1860	1881			1857/63[6]	1870/2[7]	
	2	3	4	5	6	7	8	9	10	11	12	13
		Millions				Millions of hectares				Millions of chetverts		Millions of chetverts
Upper Blackearth												
Tula	10.0	1.15	1.34	+17	3.2	2.20	2.26	+3	L.−	+1.36	+3.30	+1.61
Ryazan	9.7	1.42	1.71	+20	3.1	2.36	2.34	−1	L.+	+1.73	+1.80	+1.05
Central Agricultural Provinces												
Oryol	11.2	1.53	1.89	+24	2.9	2.57	2.90	+13	H.+	+2.42	+3.94	+1.12
Kursk	10.6	1.83	2.31	+26	3.1	3.11	3.44	+10	L.+	+1.36	+7.16	+1.23
Tambov	11.2	1.97	2.49	+26	3.5	4.0	4.20	+5	V.H.+	+0.92	+5.83	+2.18
Voronezh	6.2	1.94	2.43	+25	3.2	3.97	4.55	+15	L.+	+2.47	1.73	+0.30
Middle Volga												
Penza	0	1.18	1.38	+17	3.9	1.69	2.41	+43	L.+	+1.56	+1.07	+0.46
Vyatka	0	2.22	2.74	+23	3.2	3.69	4.76	+29				
Kazan	0	1.61	1.96	+22	3.1	2.80	3.12	+11	H.+	+1.69	+0.73	+1.26
Simbirsk	0	1.18	1.47	+25	4.0	2.03	2.53	+25	L.+	+2.46	+1.20	+0.93
South-East												
Saratov	4.3	1.69	2.11	+25	3.6	2.24	4.84	+116	V.H.+	+3.53	+2.49	+1.37
Samara	0	1.69	2.22	+31	3.6	2.15	7.53	+250	V.H.+	+3.12	+3.32	+2.53
Orenburg	0	{1.84	1.20	{+61	{3.7	{1.75	6.42	+425	{V.H.+	{+2.88	{+3.18	+0.43
Ufa	0		1.77				2.79					+0.89

Province												
Baltic Provinces												
Estland	13.5	0.31	0.38	+23	4.3	0.30	0.34	+16	N.A.	+0.04	-0.22	-0.07
Lifland	3.9	0.93	1.17	+26	4.4	0.97	1.02	+5	N.A.	+0.51	+0.44	+0.08
Courland	8.0	0.57	0.64	+12	5.0	0.61	0.68	+13	N.A.	+0.18	+0.05	+0.75
Lithuania												
Grodno	15.9	0.89	1.23	+38	2.9	1.28	1.55	+21	N.A.	+0.34	+0.25	+0.33
Vilna	10.7	0.90	1.20	+33	2.3	1.68	1.70	+1	N.A.	+0.20	+0.96*	+0.14
Kovno	13.3	1.05	1.44	+37	3.6	1.49	1.46	-2	N.A.	+0.44	-0.39	+0.49
Kiev	5.7	2.01	2.51	+25	4.3	2.91	2.91	0	N.A.	+1.17	+3.29	+1.34
Podolia	11.2	1.87	2.28	+22	3.6	2.19	2.69	+23	N.A.	+1.40	+1.20	+1.64
Kharkov	8.8	1.59	2.16	+36	3.1	2.52	3.49	+39	N.A.	+0.77	+0.65	+0.39
Poltava	4.8	1.91	2.47	+29	3.6	2.20	3.29	+50	N.A.	+1.77	+2.46	+1.22
Volynia	7.0	1.60	2.06	+29	3.2	2.42	2.66	+10	N.A.	+0.93	+0.05	+0.55
New Russia												
Don Oblast	6.2	0.95	1.77	+86	4.2	4.20	7.41	+76	N.A.	+0.73	+0.16	+2.02
Yekaterinoslav	7.4	1.20	1.41	+18	3.4	2.04	3.36	+65	N.A.	+1.58	+0.58	+1.13
Tavrida	5.7	0.61	0.96	+57	5.0	1.08	2.35	+117	N.A.	+0.57	+0.92	+1.37
Kherson	13.4	1.33	1.87	+41	3.9	3.20	3.49	+9	N.A.	+1.96	+0.60	+2.75
Bessarabia	4.2	1.03	1.42	+38	5.1	1.21	1.67	+38	N.A.	+1.32	+0.50	+1.57
Average for all provinces				+27.7	3.2			+14.2				

N.A.. Not available

Sources: [1] Bliokh, vol. I, Table XIX, 99.
[2] For 1863, Rashin, 44 and 45.
For 1882, Borkovskiy, *Ocherk sved.*, 14–19.
[3] Kovalchenko, 1959, Table 2, 68.
For more detailed annual figures, Bliokh, vol. II, appendix, Table II.
[4] Yatsunskiy, 1961, 125, 126 and 127.
Emendation of figures in *Svod. stat. sved.*, vol. I, 44 and 45.
[5] For 1851–60, Kovalchenko, 1959, appendix, Table I, 81, 82 and 83.
For 1857–63, Vilson, 108 and 109.
For 1864–6, *Voyenno-stat. sborn.*, 242 and 244.
For 1870–2, *Stat. vrem.* series 2, vol. 10, Yershov.
[6] Vilson, 109 and 110.
[7] Bliokh, vol. II, 25.
[8] Tengoborskiy, vol. I, 142.
[9] Borkovskiy, *Ocherk sved.*, 14–19.
For 1875 see Shmeyn, VII and VIII.

4. *Far north*

Isolated with no railways—low rate of population increase—some expansion in grain production—a reduction in the grain deficit—more or less self-sufficient according to Borkovskiy's figures.

5. *Upper blackearth belt*

Fairly high railway density—a low rate of population increase—little change in the cultivated area or surplus.

6. *Central agricultural region*

Fairly high railway density (except Voronezh)—rate of population increase slightly below average—some expansion in grain production—a big increase in the grain surplus.

7. *Middle Volga*

No railways—a low rate of population expansion—a marked increase in grain production

8. *The South-east*

Practically no railways—a rapid rate of population increase—a very rapid expansion in grain production—an increased surplus.

9. *Baltic provinces*

Average railway density (Estland well above, Lifland well below)—a lower than average rate of population increase—some extension of the cultivated area—a reduction in grain surplus—a largely self-sufficient area.

10. *Lithuania*

High railway density—high rate of population increase—little change in cultivated area—a slight reduction in grain surplus.

11. *Ukraine*

About average railway density—about average rate of population increase—marked expansion in cultivated area (except Kiev and Volynia)— some increase in grain surplus.

12. New Russia

Low railway density (except Kherson province)—a very rapid rate of population increase—a big expansion in the cultivated area.

With regard to column 2 of the tables, the reaction of the agricultural sector to railway construction must have involved some time-lag. Moreover, the first wave of rapid railway construction ended in 1874. Using data for railway density at the beginning of 1874 does not, therefore, give too misleading a picture.

In columns 3, 4 and 5 the percentage increases are rounded to the nearest whole number. The figure for Yekaterinoslav is clearly an underestimate as the population in 1885 was 1·79 million.

The grain yield in column 6 is an average for the years 1861–60. Comparison with yields in the 1840s, 1860s and 1870s shows that there may be a tendency to underestimate yields in the central agricultural region and the west, and to overestimate yields in New Russia and in the lower Volga region.

In columns 7, 8 and 9, in view of the predominance of grain production in total crop production, figures for the total cultivated area can be taken as a proxy for the area sown to grain. In some cases an element of distortion arises because of increased diversification in crop cultivation, for example, increased potato cultivation in the west.

The figures used here are Yatsunskiy's revision of those quoted in the *Svod Stat. Sved.*, which in their turn are based on Vilson's figures for the beginning of the 1860s and the 1881 figures derived from a survey of agricultural holdings by the Central Statistical Committee. The major emendations are a reduction in the area cultivated in the western provinces at the beginning of the 1860s and a rather higher overall increase in the total cultivated area, of 30 per cent rather than 14 per cent.

As a check on the preceding columns the trend in sowings of grain between 1857–63 and 1870–2 is indicated in column 10. The following symbols are used:

V.H. +   an increase of over 20%
H. +   an increase of over 10%
L. +   an increase of between 0% and 10%
L. −   a decrease of between 0% and 10%
H. −   a decrease of over 10%
V.H. −   a decrease of over 20%

In case these trends are statistical accidents reference has also been made to the figures for 1851–60 in Kovalchenko, 1959, and for 1864–6 in the *Voyenno-Statisticheskiy Sbornik*.

In columns 11 and 12 the size of the potential surplus is given by province for the relevant periods. There is a danger in using a three-year period as the basis for working out the potential surplus. For example, the harvest in the central agricultural region for 1870–2 was high in each of these years. The potential surplus is therefore exaggerated, particularly for Kursk and Tambov.

An asterisk denotes those provinces that were already considered to be deficit provinces in the first half of the nineteenth century.

Finally, column 13 shows the net grain exports or imports of each province, including inter-provincial movements by rail and water, but not by cart.

The harvests for 1878, 1880 and 1882 were good, bad and indifferent respectively, so that the picture presented by average net exports (+) or imports (−) is reasonably typical. Inclusion of Shmeyn's figures for 1875, a year of poor harvest, would not increase the typicality of our average figures.

Net 'exports' are slightly underestimated because they have been converted from poods into chetverts at the rate of one chetvert equals nine poods.

In this interpretation attention is concentrated, firstly, on the main grain deficit areas—the metropolitan provinces, the central industrial region and the west. Secondly, we are interested in the provinces supplying these areas, the Volga provinces and the central agricultural region. We can classify the provinces under consideration into groups.

Deficit area
1. Those which show a dramatic move away from grain

production. These are characterised by low yields, a big contraction in the cultivated area and the amount of grain sown and a large increase in the grain deficit. Moscow, St Petersburg, Tver, Smolensk, Kaluga, Vladimir and Yaroslavl provinces fall within this group.
2. Those which show some movement away from grain production. These are characterised by lower than average yields, a marked contraction in the cultivated area and/or the amount of grain sown. Kostroma, Pskov, Mogilev and Vitebsk provinces probably fall in this group. In the western provinces grain production was to a large extent replaced by potato production.
3. Those which show little change. In this group yields are average, but the other indices show a very small movement or a movement in opposing directions. Novgorod and Minsk provinces should be included in this group.

Surplus area
4. Those which show some movement away from grain production. These are characterised by about average grain yields and a tendency for the cultivated area and/or the amount of grain sown to fall. Some are also characterised by a fall in the potential surplus. A belt of provinces dividing the surplus from the deficit areas falls within this group—Nizhniy Novgorod, Ryazan, Tula and Chernigov provinces.
5. Those which show a movement towards increased grain production. These are characterised by above-average grain yields, by an increase in the cultivated area and amount of grain sown and by a tendency for the potential surplus to increase. Kursk, Oryol and Tambov provinces conform to these tendencies. The Volga provinces of Penza, Simbirsk, Kazan and Saratov, and also Voronezh province show expansion without an increase in the surplus.
6. Those which show a big increase in grain production. These are characterised by high grain yields, by big increases in the cultivated area and in grain sown, and by increases in the potential surplus. These provinces are invariably those with a very high rate of population increase reflecting new settlement—Samara, Ufa and Orenburg provinces.

The general movement towards increased regional specialisation is quite clear. Against an overall increase in the cultivated area of over 30 per cent the metropolitan provinces, the 'core' provinces in the central industrial region and some western provinces show reductions of over 15 per cent. Furthermore, particularly in the metropolitan provinces and the centre, the potential deficit rose markedly. Evidently, in this area of low grain yields the surplus population released from the land moved either to the urbanised metropolitan provinces of Moscow and St Petersburg or joined in the settlement of the new areas in the south and south-east. The ring of provinces around Moscow and along the Nikolayev railway show very low rates of population expansion. In the provinces of Smolensk and Tver, where grain yields were at their lowest and little alternative employment existed, population increase was as low as 4·4 per cent and 6·6 per cent respectively over the nineteen-year period. The provinces of Novgorod, Yaroslavl and Vladimir all show a rate of increase of just over 10 per cent. However, the metropolitan provinces show an increase of almost 40 per cent, despite the fact that their natural rate of increase was probably under 10 per cent.[1] This means that Moscow and St Petersburg provinces absorbed at least three-quarters of a million people from the neighbouring provinces.[2]

---

[1] The natural rate of increase of population in a rural society is probably related to the supply of cultivable land. This is likely to be true in Russia where land was often redistributed every few years according to the number of consumers or male workers in a family. In the provinces of Kaluga, Tula and Ryazan population increased by between 16·5 and 20·5 per cent. While these provinces almost certainly lost population to Moscow, their natural rate of increase was probably slightly higher than in the rural areas of Moscow province. Vologda and Kostroma provinces, out of direct reach of the pull of both Moscow and the new areas in the south, show an increase of between 18 and 20 per cent. However, the advantages of having children were much less in urban areas and the death rate much higher. According to Yatsunskiy, between 1867 and 1897 the natural rate of increase in the provinces under discussion was about 1 per cent per annum, but it was as low as 0·45 per cent in Moscow province and 0·12 per cent in St Petersburg province. On the basis of these figures a rate of 10 per cent for the period under discussion might even be an exaggeration.

[2] Yatsunskiy's figures for the period 1867–97 suggest a net migration into Moscow and St Petersburg provinces of 1·3 million. (See Yatsunskiy, 1957, 210.)

This area and the region of the Upper and Middle Volga must also have supplied settlers to the newer areas of settlement in Samara, Orenburg and New Russia.

Certainly the trend in favour of increased specialisation had begun at a much earlier date. The logic of our argument would make any improvement in transportation instrumental in promoting specialisation, as for example canal construction at the beginning of the nineteenth century. Our argument is contained in two propositions. Firstly, the effect of railway construction was to accelerate the rate of regional specialisation, and to raise the average grain yield through this factor rather than through improving methods of production.[1] Secondly, railway construction determined the exact pattern of regional specialisation. Our evidence in support of the first proposition consisted in showing a dramatic increase in and concentration of the marketed surplus and in referring to an acceleration in the rate of urbanisation. The second proposition requires discussion.

Reference to Tables 5 and 6 shows that, with the exception of new areas of settlement, the biggest changes occurred on a line running along the railways linking Saratov province with Tambov, Tambov with Moscow and Moscow with St Petersburg. Moscow, Tver and St Petersburg provinces all experience a dramatic movement away from grain production. Tambov and Saratov provinces experienced a significant movement towards grain production. Other provinces crossed by railway lines radiating from Moscow also experienced similar changes, Smolensk, Vladimir and Yaroslavl as consuming and Oryol and Kursk as producing provinces. The latter two provinces

---

[1] According to Nifontov the average net harvest was 145 million chetverts in 1858–62 and 174 million in 1879–83, an increase of 31 per cent. Yatsunskiy's figures show an increase in the arable area of 30 per cent. Even if we allow that this latter figure is exaggerated and that there was some increase in the diversification of crops, little room is left for an increase in yields. However, we must allow for other factors:

(*i*) the elimination of marginal land yielding low returns and the cultivation of new areas of high yield;
(*ii*) some reduction in the amount of land left fallow;
(*iii*) some reduction in the degree of underestimation of the total harvest

It is clear that proper account of all these factors would leave no room for an increase in yields on already settled land through improved technology or organisation.

were of course also affected by the growth of exports through the Baltic ports.

THE RAILWAYS AND DOMESTIC GRAIN CONSUMPTION IN THE MAIN CENTRES

We turn now to the third part of the paper in which changes in the size and supply pattern of the main domestic grain markets, Moscow and St Petersburg, are analysed. As Borkovskiy's figures indicate, the main domestic markets for grain lay in the central non-blackearth region and the north, whereas the main supplying areas were in the central blackearth region and along the lower Volga. The deficit in the centre amounts to 38·3 million poods of grain, in the north to 30·1 million.[1] At the heart of the two consuming areas were Moscow and St Petersburg. These were the two most important single markets for grain.

Table 7 shows the level of grain consumption in Moscow and its change over a period of time. The figures show an apparent fall in total consumption from about 25 million poods to about 18 million in the second half of the 1860s and a rise to 26 million in the mid-1870s. At first glance these figures appear unlikely. Most evidence suggests that consumption per head was reasonably stable during this period. Therefore, since the population of Moscow was rising rapidly during this period we would expect that total consumption would also rise, probably by something in excess of 5 million poods. Judging by the figures for earlier years, the figure for 1861 seems reasonable. There are several reasons why the picture presented may be fairly accurate. Firstly, the combined impact of the Emancipation and of railway construction may have initially disrupted the whole system of grain supply. Elements of disruption may have persisted until 1866 and 1867. Secondly, the faster and more regular transport by rail reduced the need for large stocks to be held in Moscow. During the 1860s there may have been significant consumption out of stocks. Thirdly, the appearance each winter of a large

---

[1] See Borkovskiy, 1888, 7.

number of peasant carters with their horses must have involved considerable consumption of grain.[1] How much of this was

TABLE 7

ANNUAL CONSUMPTION OF GRAIN IN MOSCOW
(in millions of poods)

Year	Water	Cart			Rail			Total consumption
		Total imports	Re-exports	Retained imports	Total imports	Re-exports	Retained exports	
1789[1]	0.8	4.2		3.2				4.0
1810[1]	6.3							
1833[2]	3.9	9.0		8.0				11.9
1838[3]	9.9		⎫ 1 million					
1840[2]	9.3	9.0	⎬	8.0				17.3
1847[2]	6.5	15.2	⎭	14.2				20.7
1859–62[4]	10.1							
1861[5]	11.4	14.0		13.0				24.4
1866[6]	4.4	7.4	0.5	6.9	15.3	13.0	2.3	13.5 ⎫
1867[6]	5.5	3.6	0.5	3.1	21.3	13.0	8.3	16.9 ⎪
1868[6]	1.6	2.0	0.4	1.6	38.5	20.6	17.9	21.1 ⎬ av. 18.2
1869[6]	0.9	12.5	0.4	1.1	37.4	17.6	19.8	21.8 ⎪
1870[6]	0.8	1.0	0.3	0.7	29.2	12.9	16.3	17.9 ⎭
1874[7]					50.0	26.3	23.7 ⎫	
1875[8]	1 million						27.4 ⎬ av. 24.8 av. 26
1876[9]							23.4 ⎭	

Note: The figures for re-export by cart before 1866 and for import by cart and water after 1866 are rough estimates.

Sources: [1] Archival sources quoted by Kovalchenko, 1963, 474.
[2] *Voyenno-stat* . . . vyp. I, appendix, Table 14 of volume 4.
[3] *Vidy* . . .
[4] *Stat. vrem.* vyp. I, razd. II.
[5] Chaslavskiy, Part I, 178 and 217.
[6] Chaslavskiy, Part I, 217.
[7] Bliokh, vol. II, 64, 65 and 66.
[8] Shmeyn, 48, 49, 50 and 51.
[9] *Stat. sborn.*, vyp. 3, 16 and 17.

included in the figures for import of grain is not clear. Fourthly, the construction of a network of railways radiating from

---

[1] According to A. Zablotskiy, summer carting occupied 800 000 men and winter carting more than 3 million. A large proportion of these would have been occupied in the central region. See Zablotskiy, 34 and 35.

Moscow allowed large areas to be supplied by rail directly rather than through Moscow. This was particularly true for the industrial areas to the north and east of Moscow. Moscow, therefore, declined in importance as a grain market for the surrounding areas. In other words, the number of people provisioned by the total consumption figures was reduced. It is possible that during the 1860s and 1870s this factor alone fully offset the impact of rising population. Reference to the whole of Moscow province would eliminate this factor. In fact, as previously indicated, the level of consumption in the province as a whole seems to have risen very fast.

All the four factors described above would help to explain the large fall in consumption in the 1860s. The first two are once-and-for-all effects worked out in a few years. The last two are non-cumulative factors. The impact of rising population and increasing urbanisation would, therefore, eventually reassert itself in raising consumption, as it appears to do in the 1870s.

The table also shows the rapid decrease in importance of waterborne and carting traffic. Furthermore, the intial impact of the railways was to change the pattern of supplying Moscow's domestic needs. In the 1870s a big expansion in re-exports took place.

For St Petersburg the figures show a steadily rising trend (see Table 8). Over the ten-year period from the mid-1860s to the mid-1870s consumption rose by only one and a half million poods. Since, with the building of new railways into Finland and along the Baltic coast, the area supplied from St Petersburg must also have shrunk, but with a smaller impact owing to sparser population, the shortfall on what might be expected from population figures alone, say 3 million poods, is easily explained. The earlier figures also indicate a practically stable level of consumption during the 1840s and 1850s. As St Petersburg was never as important a grain market as Moscow the fall in consumption discovered for Moscow would not be expected for St Petersburg. Again, it should be pointed out that a stable level of consumption is not inconsistent with an accelerated rate of urbanisation which would be reflected in the consumption figures for St Petersburg province as a whole.

TABLE 8

ANNUAL CONSUMPTION OF GRAIN IN ST PETERSBURG

(in millions of poods)

Year	By water	Imports By cart	By rail	Total	Exports	Consumption
1775[2]	10·4					10·4
1781[1]	9·2					9·2
1786–8[1]	10·4					10·4
1796[2]	14·8					14·8
1801–6[1]	12·8					12·8
1811[1]	16·5					16·5
1813[1]	15·5					15·5
1818[1]	19·6					19·6
1837–9[3]	24·6	1·0		25·6	1·7	22·9
1854[4]	33·5	1·0		34·5	0·1	34·4
1854[1]	32·8	1·0		33·8	0·1	33·7
1859–62[5]	26·5	0·5	3·6	30·6	7·0	23·6
1864–7[6]	30·9	0·4	10·2	41·5	18·0	23·5
1874[7]	28·5		36·5	65·0	45·9	19·1
1875[8]	25·0		36·9	61·9	40·5	21·4
1876[9]	35·2		48·7	83·9	56·7	27·2
1877[9]	30·2		73·1	103·3	78·3	25·0
1878[9]	21·8		51·6	73·4	40·5	32·9
(1874–8 av.	28·1		48·4	76·0	52·2	25·1)

Notes: (i) Before 1837–9 imports by cart and exports must both have been positive but we assume that they cancel each other.
(ii) The export figure for 1837–9 is the average export of the four main grains over the period 1824–47.
(iii) Obviously the figures for 1854 reflect distortions arising from the Crimean War.

Sources: [1] Archival sources quoted by Kovalchenko, 1963, 475.
[2] Rubinshteyn, 406.
[3] *Vidy* ...
[4] *Obzor* ...
[5] *Stat. vrem.*, vyp. I. *razd.* II.
[6] Borkovskiy, 1872, appendix IX.
[7] Bliokh, vol. II. 64, 65 and 66.
[8] Shmeyn, 72, 73, 74 and 75.
[9] *Stat. sborn.*, vyp. 3.

It is clear from the table that total movements of grain into St Petersburg increased significantly, almost doubling. The increase was mostly explained by a dramatic increase in the export of grain through St Petersburg and Kronstadt, amounting almost to a tripling.[1] Furthermore, movements of grain by water declined slightly, whereas movements by rail increased almost five times.

THE PATTERN OF SUPPLY

Changes in the pattern of supply were far greater than in total consumption of the two cities. Traditionally Moscow was supplied overland and by water from the blackearth provinces to the south—from southern Tula and Ryazan, eastern Oryol and the western and northern areas of Voronezh, from the western fringes of Saratov and Penza and, finally, from Kursk and Tambov provinces. St Petersburg was supplied from the provinces on the lower Volga by way of the three canal systems linking the Volga with the Neva river, from Saratov, Samara, Simbirsk, Kazan and Penza provinces. Moscow was largely supplied by cart, St Petersburg predominantly by water.

Overland the main supply route to Moscow was the Ryazan or Astrakhan tract, which passed through Kozlov, Ryazhsk, Ryazan and Kolomna. This route was closely rivalled by a second tract running west from Voronezh and Yelets through Lebedyan, Dankov, Mikhaylov and Zaraysk to Kolomna. Another tract ran from Voronezh through Yelets to Yefremov where it joined the road from Livna. From there branches went to Yepifan, Bogoroditsk and Tula. The former two met again at Venyev but diverged once more, one to Zaraysk and Kolomna, the other to Kashira and Moscow. The latter was joined by the route from west Ryazan. Further west the Muravka tract ran from the southern part of Kursk province through its eastern half to Livna and Tula. Finally another

---

[1] The average export of the four main grains through St Petersburg was as follows, in millions of poods:

1856–60	11·4	1861–5	6·8
1866–70	18·0	1871–5	32·3
1876–80	49·7		

There are clear jumps to new plateaux in 1866 and 1871, following shortly after the completion of the Ryazan–Kozlov and Rybinsk–Bologoye lines respectively. *Source:* Pokrovskiy, 36–9.

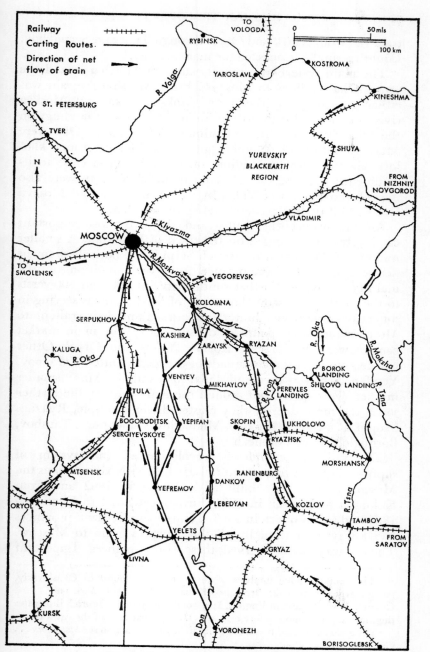

Fig. 3. Grain supply routes of Moscow

tract ran from Kursk and Oryol through Tula and Serpukhov. Countless smaller tracks fed the main arteries.[1]

The main waterborne routes ran from Morshansk and the upper Oka landings to Moscow. From Morshansk grain was either moved along the Tsna, Moksha, Oka and Moskva rivers or more directly to the Shilovo and Borok landings on the Oka river. The main landings on the upper Oka were Mtsensk and Oryol. Another route ran to Moscow from Ukholovo via the Perevles landings on the river Pron close to its confluence with the Oka. Small quantities of grain were also moved from the lower Volga by way of the Oka and Moskva rivers and from the Ryazan landings on the Oka.

All the waterborne routes and the two most important carting tracts converged on the town of Kolomna. A smaller number of routes passed through Serpukhov. Both these towns derived their chief importance from their position on the main grain routes. Further south, between 300 and 400 versts to the south of Moscow, lay a belt of market towns serving in some cases a very wide hinterland and sending their grain on to Moscow and other points of consumption. The main market centres were Morshansk, Kozlov, Yelets and Oryol. Other markets stood at the crossroads of the grain routes. Yefremov, Yepifan, Sergiyevskoye and Tula, for example. Many smaller markets were spread out around the main centres, the further south the more densely they clustered: for example, Ryazan, Skopin, Ryazhsk, Ukholovo, Mikhaylov, Ranenburg, Tambov, Borisoglebsk and Voronezh.

The railways completely revolutionised this system of supply.[2] The most important railway link with Moscow was the Moscow-Ryazan line built to Kolomna in 1862 and from Kolomna to Ryazan in 1864, further extended to Kozlov in 1866 and to Voronezh in 1868. Tambov and Ryazan provinces had always sent virtually all their surplus grain to Moscow. The railway mainly served these two provinces. Important

---

[1] The most detailed analysis of these routes is found in Chaslavskiy, particularly Chapter 2. See also Kovalchenko, 1959 *Krest'yane* . . . Chapter 1, 54–62, and in English, Haywood, Chapter I. Figure 3 illustrates the main supply routes before and after the construction of the railways.

[2] For a chronological breakdown of railway construction see *Stat. sborn* . . . vyp. 2.

feeder lines linked the railway with Morshansk in 1867, Tambov and Borisoglebsk in 1869, Umet and Filonovo in 1870.

The Moscow–Ryazan railway and its extensions ran parallel with the Ryazan tract. As it was pushed further south grain was diverted from the carting route to the railhead. Grain was also diverted from the tract running from Voronezh through Kolomna. The other tracts also became insignificant with the opening of the Moscow–Kursk line between 1866 and 1868. According to the information of the Moscow duma, the amount of grain entering the market in Bolotnaya Square, where nearly all grain transported to Moscow by cart was sold, fell from 13–15 million poods before the advent of the railway to just under 8 million poods in 1866, 1·75 in 1868 and less than 0·8 million in 1870. By the 1870s the carting trade in grain had virtually ceased to exist.[1]

A similar experience confronted the waterborne trade. The construction of the Moscow–Kolomna line diverted grain from the Moskva river. The Ryazan landings were reduced to insignificance by the extension of the line, first to Ryazan itself, then to Kozlov. The building of the Ryazhsk–Morshansk line had the same effect on despatches from the Tsna, Shilovo and Borok landings, particularly therefore from Morshansk itself, and from the Perevles landings. Movement of grain from the Upper Oka landings to Moscow also virtually ceased with the completion of the line to Kursk. Shippings from the lower Volga to Moscow had already been deflected to the Moscow–Nizhniy Novgorod line, completed as early as 1862. The overall impact on the waterborne import of grain into Moscow was to reduce it from a figure of 11·5 million poods at the beginning of the 1860s to a lower plateau of 7·25 million in 1863 and 1864, 4·5–5·5 million in 1866 and 1867 and less than one million by the end of the decade.[2]

The impact of the construction of the Moscow–Ryazan–Kozlov–Voronezh line, however, was not limited to a change in the method of transport used in supplying Moscow. It now became possible, using this line and the Nikolayev railway, to supply St Petersburg at a cost competitive with supply from

---

[1] Chaslavskiy, Part 1, 178.   [2] Ibid.

the lower Volga provinces, even after the introduction of steamships on the Volga. The result is reflected in the figures for grain movements along the Nikolayev line. At the beginning of the 1860s more grain arrived in Moscow on this line than was despatched.[1] Grain despatched amounted to less than one million poods. In 1862 dispatches for the first time exceeds arrivals. By 1864 arrivals were negligible.[2] Despatches became significant in 1865, at 8 million poods, and continued to rise, to just under 17 million in 1869[2] and about 24 million in the mid-1870s.[3] A sharp jump in exports of grain through St Petersburg in 1866 is probably accounted for by the construction of the Moscow–Ryazan–Kozlov line.

The surplus of grain imported into Moscow along the Moscow–Ryazan line reflected, therefore, both the needs of the local population and St Petersburg's domestic and export demands. The figure grew rapidly and far exceeded movement along any of the other lines radiating from Moscow. In 1866 just over 14 million poods were received, in 1869 just over 27 million[4] and by 1876 over 38 million.[5] This last figure is almost three times the imports of grain by other lines. The reason for this is quite clear. The only other line directly serving the blackearth region was linked, in 1868, with Riga, the major export outlet on the Baltic. It became more profitable to send surplus grain, particularly from Oryol and Kursk provinces, to the Baltic, and even parts of Tula and Tambov provinces and north Voronezh fell, after 1870, within the catchment area of the Baltic ports.[6]

The supply system of St Petersburg was much simpler.[7]

---

[1] Vilson, 158.   [2] Chaslavskiy, Part I, 180.
[3] Bliokh, vol. II, 67 and 68; *Stat. sborn* . . . , vyp. 3, 20 and 21 (*Svedeniya* . . .).
[4] Chaslavskiy, Part 1, 217.
[5] *Stat. sborn* . . . , vyp. 3, 20 and 21.
[6] Chaslavskiy, Part II, Chapter 1.
[7] Borkovskiy's work gives detailed information on the canal systems, Shubin on movement along the Volga. A good deal of this information is summarised in *Obzor* . . . , 1893.

In English there are two reasonably good surveys of the water routes from the lower Volga to St Petersburg, in Haywood, Chapter 1, Blackwell, Part IV, Chapter II.

Figure 4 illustrates the main supply routes before and after construction of the railways.

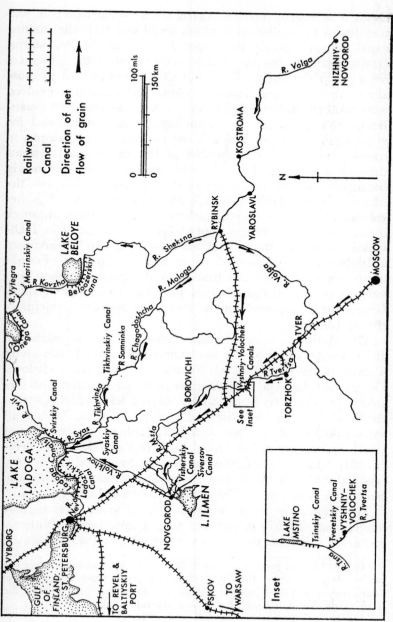

Fig. 4. Grain supply routes of St Petersburg

Grain was moved up the Volga to Rybinsk where it was transferred to smaller craft which could negotiate the shallow canal systems linking the upper Volga with St Petersburg. Grain was fed into the Volga network from the provinces adjacent to the Kama, Sura and Oka rivers as well as those along the lower and middle Volga. The distances involved were prodigious. From Astrakhan to Rybinsk was 2565 versts. Before the advent of the steamship it was not unusual for grain from the lower Volga to be two navigation seasons in transit; before the construction of the three canal systems this was the norm. Two of these canal systems were newly constructed at the beginning of the nineteenth century—the Mariinskiy and Tikhvinskiy systems. The third—the Vyshnevolotskaya system—was refurbished at the same time, although it had been begun in the time of Peter the Great. The Tikhvinskiy route was more direct than the other two but involved a higher cost because of its greater shallowness. Freight carried on this route was limited in the main to either high value or perishable commodities, for which speed was essential. Grain was transported on the two other systems, principally the Mariinskiy.[1]

The Mariinskiy system involved transport over a distance of 1073 versts.[2] This was the most northerly route. It ran along a series of rivers and canals linking the three lakes—Beloye, Onega and Ladoga. The four sections before, after and between the lakes were linked by the rivers Sheksna and Kovzha, the river Vytegra, the Svir and finally the Neva. In fact all three routes converged south of Lake Ladoga. The sections of canal constructed either linked the rivers or lakes—the Beloozerskiy and the Mariinskiy canals—or by-passed the stormy lakes, the Onega canal and the series of canals linking the Svir, Syas, Volkhov and Neva rivers south of Lake Ladoga.

The Vyshnevolotskaya system, at 1187 versts, was slightly longer than the Mariinskiy and took a much more southerly route. On this route the Volga was left only at Tver. Again the

---

[1] Of 30·9 million poods reaching St Petersburg by boat on average per year during the period 1866–8, 21·2 million poods were moved on the Mariinskiy system, 7·7 on the Vyshnevolotskaya and only 2·0 on the Tikhvinskiy. See Borkovskiy, *Izsled* . . . , appendix, Table IX.

[2] See the map for further information on these routes.

route consisted of a whole series of interlocking lakes—Lakes Mstino and Ilmen—rivers—the Tvertsa, Tsna, Msta and Volkhov rivers—and canals, the Tveretskiy, Tsninskiy and Siversov canals. This route converged on the first at Novaya Ladoga.

Finally, the Tikhvinskiy more or less bisected the other two routes and covered a distance of only 847 versts. However, a more intricate pattern of small rivers and lakes made the passage of large craft difficult on this route. The main river links were the Mologa, the Tikhvinka and Syas rivers. The Tikhvinskiy canal linked the basin of the first with that of the latter two.

The impact of railway construction on the supply system for St Petersburg was far less dramatic in the displacement of existing methods of transport than it was for Moscow. However, as we have seen, on grain imports the impact was far more impressive because of the large increase in the export of grain through St Petersburg and Kronstadt.

The opening up of the central agricultural region to railway transport led to a big increase of traffic on the Nikolayev railway. Receipts of grain at St Petersburg rose from 3·6 million poods on average between 1860 and 1863[1] to 10·2 million poods between 1864 and 1867,[2] or from 2·6 million to 15·9 million at the terminal dates, 1860 and 1867. The next significant event was the opening of the Rybinsk–Bologoye railway in 1870 which allowed the three canal systems linking the Volga with St Petersburg to be by-passed and fed even more grain on to the Nikolayev railway, incidentally giving back the Volga provinces some of their cost advantage. Arrivals on the Nikolayev railway averaged 50 million poods in the period 1874–8.[3] Shipments on the Rybinsk–Bologoye railway built up very quickly, reaching 18·5 million poods in 1874 and 35 million in 1876, which exceeded despatches from Moscow on the Nikolayev line by as much as 12 million poods.[4]

The impact of railway construction on movements of grain

---

[1] Vilson, 59.
[2] Borkovskiy, *Izsled* . . ., appendix, Table IX.
[3] Bliokh, vol. II, 72 and 73; Shmeyn, 72 and 73; *Stat. sborn* . . ., vyp. 3, 16 and 17 (*Svedeniya* . . .); *Stat. sborn* . . ., vyp. 3, 3 and 6 (*Pribytiye* . . .).
[4] Bliokh, vol. II, 72 and 73; *Stat. sborn* . . ., vyp. 3, 16 and 17.

from the lower Volga was probably marginally negative, in the sense that they were lower than they would have been in the absence of the railways. The following figures help us to clarify the situation:

TABLE 9

NET EXPORTS OF GRAIN FROM THE VOLGA PROVINCES
(in millions of chetverts)

Province	1863 by water	1875 by water	1875 by water and rail	1878, 1880 and 1882 (by both)
Kazan	1·9	2·4	2·4	1·3
Simbirsk	0·7	1·42	1·47	0·9
Samara	2·8	3·3	3·3	2·5
Saratov	1·5	0·8	2·0	1·4
TOTAL	6·9	7·9	9·1	6·1

Note: The total harvest was good in 1863 and bad in 1875 but the yields in the provinces along the Volga were about the same in the two years. Since the demand for Volga grain would have been higher in 1875 one would expect net exports to have risen on this account. The average yield for the years 1878, 1880 and 1882 was probably somewhat lower than for the other two years.

Sources: 1963 figures—Rudanovskiy, No. 11, 18.
1875 figures—Shmeyn, VII.
1878, 1880 and 1882 figures—Borkovskiy, *Ocherk* . . . , 14 and 15.

It certainly looks as if there was very little, if any, increase in net exports from the Volga provinces. Meanwhile, over the period 1864–7 to 1874–8 the total movement of grain to St Petersburg had risen from 4·6 million chetverts to 8·4 million. In other words, in the early period the Volga more than met the needs of St Petersburg. It also met the needs of much of the upper Volga basin, the central industrial region and Moscow. However, by 1875 it barely met the needs of St Petersburg[1] and by the end of the 1870s it could not do even this.

The effect of railway construction on movement on the Volga is indicated by the figures for Saratov. While movement by

[1] See Table 8.

water from Saratov province declined from 1·5 million chetverts to 0·8 million between 1863 and 1875, rail movement accounted for 1·2 million at the latter date. Thus, despite the disappearance of *burlachestvo* under the competition of the steamboat, it appears that the shipment of grain by boat could not compete.

THE RELATIVE ADVANTAGES OF THE DIFFERENT TRANSPORT MEDIA

To illustrate the relative advantages of transport by rail, water and cart, three routes are considered: Saratov–St Petersburg, Morshansk–Moscow and Kozlov–Moscow. Table 10 summarises the data on distance, time taken and cost of rail or waterborne transport over these routes.

The speed of movement by the different modes of transport varied greatly. At best freight moved by cart on chaussées or good winter roads averaged no more than 50 versts per day. On the earth roads of the central region it could do no more than 30.[1] During the *rasputitsa* or dusty summer months movement was even slower.

By boat the speed varied according to water depth.[2] Steamboats were introduced at first on the Volga, but on the shallower rivers of the central provinces or the canal systems of the north traditional methods continued. On the Volga the average speed of a tug-boat dragging 2–3 loaded barges against the current was 3–5 versts per hour, or 50–100 versts per day. So-called 'light' tug-boats, pulling only one barge, reached 7–8 versts. The use of a steam capstan gave an average speed of 30–50 versts per day, a capstan turned by horses 20–30 versts per day. Other methods, in particular the human traction of the burlaks, managed only 1 verst per hour or 15 a day. On the Mariinskiy canal system 20 versts per day was the best to be hoped for. On the shallow rivers of central Russia even less. On the railways, however, 300 versts per day was considered usual.[3]

---

[1] Chuprov, vol. I, 12; Golovachev, 35.
[2] Borkovskiy, *Torgovoye* . . . , 11–13.
[3] Golovachev, 36.

TABLE 10

THE RELATIVE ADVANTAGES OF TRANSPORT BY RAIL,
WATER AND CART IN 1872

Route	Distance (in versts)	Time taken	Cost (kopecks per pood)	
1. Saratov—St Petersburg				
(a) By rail:				
Saratov–Kozlov	421	15 days	Wheat flour	41·2
Kozlov–Ryazan	198		Grey grains	
Ryazan–Moscow	196		and grits	29·4[1]
Moscow–St Petersburg	604			
	1419			
(b) By water:				
Saratov–Rybinsk	1745	3 months		6*
Mariinskiy canal system	1073			8–9
	2818			14–15
(c) By rail and water:		4 weeks		
Saratov–Rybinsk	1745			6
Rybinsk–Bologoye	280 ⎫			
Bologoye–St Petersburg	245 ⎭			11·4†
	2270			17·4
2. Morshansk–Moscow				
(a) By rail:				
Morshansk–Ryazhsk	121	1 week	(Rye, oats,	
Ryazhsk–Moscow	284		wheat and flour)	14·8[2]†
	405			
(b) By water:				
Tsna, Moksha, Oka and Moskva rivers	1200	80 days		10–12·5
3. Kozlov–Moscow				
(a) By rail:				
Kozlov–Ryazan	198	just under one week	(Wheat, oats rye)	6·6[3]
Ryazan–Moscow	196			5·14
	395		Loading and unloading	1·0
				12·74[3]
(b) By cart:		about two weeks		
about some distance				26

[Notes to Table opposite

IMPACT OF RUSSIAN RAILWAY CONSTRUCTION

Of course, the time taken by water or cart varied greatly according to climatic conditions and 'acts of God'. Rail transport showed much less variability. Furthermore, the water routes were closed by ice for significant periods of the year. The rivers and canals were open for the following periods during the years 1875–7.[1]

	Months open	Closed on average
On the Volga:		
At Samara	III–IV—XI	3 months
Rybinsk	III–IV—X–XI	4 months
St Petersburg	IV—X–XII	5 months
On the Mariinskiy Canal System		
At Byt Gor	IV—X–XI	6 months

[1] *Stat. sborn.*, vyp 1: *Vskrytiye i zamerzaniye rek, ozer i kanalov.*

NOTES TO TABLE 10

* Includes cost of loading and unloading.

† In the late 1860s the cost of water transport from Saratov to Rybinsk was as high as 10 kopecks.

Note: Naturally, since the cost of transport varied both with the method of traction and the time of the year, there is no such thing as a rail cost, a water cost or a carting cost. The figures in the table represent the annual average cost for waterborne and overland transport and the 'ordinary' cost for rail movement. On the Volga we take the cost for a heavy tug-boat. The water tariff varied according to the 'race'; it was at its highest for the first race. The tariff for fast, light tugboats was 50–70 per cent higher than for the heavy tug-boats, for capstans 10–20 per cent less and for other methods, e.g. human traction, 20–30 per cent less. The basic rail freight tariff between Morshansk and Moscow was lowered during the navigation season (March to June) from 1/30 kopeck per pood-verst to 1/40 kopeck. This was done on other lines such as the Nikolayev. Overland the lowest cost would be incurred for winter movement in the central region on a good chaussée.

Account is taken of loading and unloading in the calculation both of time taken and of cost. According to Golovachev, 4–6 days were needed for loading and unloading on the railways.

The distances quoted are taken from *Stat. Sborn.*, vyp. 2, and *Obzor* . . .

The time taken has been worked out from average speeds cited in the text. These results have been checked against times given in the work of Borkovskiy, Chaslavskiy and Chuprov.

Sources: [1] Chaslavskiy, Part 1, 38–39; Borkovskiy, *Torgovoye* . . . , 17–19, *Puti* . . . , 7 and 19 (for earlier years).
[2] Chaslavskiy, Part 1, 80–82.
[3] Chaslavskiy, Part 1, 33; Chuprov, vol. I, 14–15.

The same was true of the other two canal systems.

On the Tsna		
At Morshansk	III—X–XI	5 months
On the Oka		
At Kolomna	III–IV—XI	4 months

The railways had the great advantage of being usable all the year round.

Reference to Table 10 shows that it was cost competition which undermined the carting trade. However, this does not appear to be the case with waterborne transport. In the central region cost differences were insignificant. Quite clearly the enormous time saving must have been decisive. The Volga waterway also appears to have been by far the cheaper route from the lower Volga provinces to St Petersburg. Even so, as we have seen, grain transport built up dramatically on the Nikolayev and Rybinsk–Bologoye lines and water tariffs declined greatly under the impact of railway competition.[1] Once again the massive time savings must have been decisive.

CONCLUSIONS

The propositions advanced in this paper have not been conclusively proved. In the absence of reliable and comprehensive statistical support we can only rely on the internal consistency of the propositions and their conformity to the fragmentary evidence available. Only a circumstantial case has been, and probably can be, established.

There are also elements of over-simplification in the paper. All grains have been lumped together. However, if we considered each type of grain separately we would probably find the tendency to increased regional specialisation even more pronounced. In particular the expansion of oats' production in the more northerly provinces conceals a contraction in the production of other grains, greater than our aggregate figures indicate. Classification by province also conceals increased intra-province specialisation.

---

[1] See Borkovskiy, *Torgovoye* . . . , 17 and 18.

The following propositions were advanced in this paper. Over the two decades of the 1860s and 1870s there was a big increase in the proportion of grain output marketed. Most of the increase was accounted for by an increase in exports. However, there was some expansion of the domestic market for grain. The construction of the railways was the immediate cause of increased marketings through the release of an already existing potential surplus of grain. The impact of railway construction on existing methods of production was insignificant. However, improved communications did encourage increased regional specialisation, that is, a concentration of grain production on high-yielding provinces. The pattern of grain supply to the main centres of consumption was dramatically changed, particularly in the case of Moscow. The change was occasioned, not so much by the lower costs of railway transport, as by the time savings involved.

Some even more tentative deductions can be made from these propositions. Two developments—the displacement of labour-intensive means of transport by more capital-intensive means and the freeing of agricultural labour through increased regional specialisation in grain production—both favoured the creation of a surplus pool of labour available for 'industrial' activities. An accelerated rate of urbanisation suggests that at least some of this pool was absorbed into the towns. In so far as agricultural prices were rising faster than other prices and the burden of taxation was not increasing, increased grain marketings would indicate increased rural purchasing power for the products of industrial activity.

BIBLIOGRAPHY

Bilimovich, A. D., *Tovarnoye dvizheniye na russkikh zheleznykh dorogakh*, St Petersburg, 1902

Blackwell, W. L., *The beginning of Russia's industrialization, 1800–60*, Princeton, New Jersey, 1968

Bliokh, I. S., *Vliyaniye zheleznykh dorog na ekonomicheskoye sostoyaniye Rossii*, St Petersburg, 1878

Blum, J., *Lord and peasant in Russia from the ninth to the nineteenth century*, Princeton, New Jersey, 1961

Borkovskiy, I., *Izsledovaniye khlebnoy torgovli v verkhne—volzhskom basseyne*, St Petersburg, 1872

Borkovskiy, I., *Ocherk svedeniy o vyvoze, vvoze i tsenakh otnositel'no khlebov rzhanovo, pshenichnavo, ovsa i yachmenya po guberniyam i voyennym okrugam yevropeyskoy Rossii*, St Petersburg, 1888

Borkovskiy, I., *Puti i sposoby perevozki gruzov s nizovykh pristaney reki Volgi k Petersburg*, St Petersburg, 1868

Borkovskiy, I., *Torgovoye dvizheniye po volzhsko—mariinskomu vodnomu puti*, St Petersburg, 1874

Chaslavskiy, V., *Khlebnaya torgovlya v tsentral'nom rayone Rossii*, 2 parts, St Petersburg, 1873

Chubinskiy, *O sostoyanii khlebnoy torgovli v severnom rayone*, St Petersburg, 1876

Chuprov, A. I., *Zheleznodorozhnoye khozyaystvo*, 2 volumes, St Petersburg, 1878

Fedorov, M. P., *Obzor mezhdunarodnoy khlebnoy torgovli*, St Petersburg, 1889

Goldsmith, R. W., The economic growth of Russia, 1860–1913, in *Economic Development and Cultural Change*, vol. 9, Part II, 1961

Golovachev, E., *Ob ustroystve zemskikh dorog*, Kiev, 1870

Harris, C. D., *Cities of the Soviet Union*, Chicago, 1970

Haywood, R. M., *The beginnings of railway development in Russia in the reign of Nicholas I, 1835–1842*, Durham, North Carolina, 1969

Khromov, P. A., *Ekonomicheskoye razvitiye Rossii v XIX–XX vekakh*, Moscow, 1950

Kislinskiy, N. A., *Nasha zheleznodorozhnaya politika po dokumentam arkhiva komiteta ministrov*, 4 volumes, St Petersburg, 1902

Koval'chenko, I. D., *Dinamika urovnya zemledel'cheskovo proizvodstva Rossii v pervoy polovine XIX v.* (*Istoriya SSR*, 1959, No. 1), Moscow, 1959

Koval'chenko, I. D., *Krest'yane i krepostnoye khozyaystvo ryazanskoy i tambovskoy guberniy v pervoy polovine XIX veka*. Moscow, 1959

Koval'chenko, I. D., *O tovarnosti zemledeliya v Rossii v pervoy polovine XIX v.* in *Yezhegodnik po agrarnoy istorii vostochnoy yevropy*, 1963 g., Vilnyus, 1964

Lenin, V. I., On the so-called market question, in *Collected Works*, vol. I, Moscow, 1960

Lenin, V. I., The development of capitalism in Russia, in *Collected Works*, vol. II, Moscow, 1960

Lyakhovskiy, V. M., *Zheleznodorozhnye perevozki i razvitiye rynka* (K istorii ryazansko—kozlovskoy dorogi 1860–1870–e gody), *Vestnik Moskovskogo Universiteta (Istoriya)*, 1963, No. 4

Lyashchenko, P. I., *Istoriya narodnogo khozyaystva SSR*, Moscow, 1948

Lyashchenko, P. I., *Ocherk agrarnoy evolutsii Rossii*, 2 volumes, St Petersburg, 1908

Nifontov, A. S., Khozyaystvennaya kon"yunktura v Rossii vo vtoroy polovine XIX veka, *Istoriya SSR*, No. 2, 1971

Obzor vnutrennevo sudokhodstva yevropeyskoy Rossii za 1854 god., St Petersburg, 1855

Pokrovskiy, V. I. (Ed.), *Sbornik svedeniy po istorii i statistike vneshney torgovli Rossii*, St Petersburg, 1902

Pribytiye glavneyshikh tovarov v St Petersburg po reke Neve i zheleznykh dorogam v 1877, *Stat. Sborn.*, Vyp. 3

IMPACT OF RUSSIAN RAILWAY CONSTRUCTION 45

Pribytiye glavneyshikh tovarov v St Petersburg po reke Neve i zheleznykh dorogam v 1878, *Stat. Sborn.*, Vyp. 3

Radtsig, A., *Vliyaniye zheleznykh dorog na sel'skoye khozyaystvo, promyshlennost i torgovlyu*, St Petersburg, 1896

Rashin, A. G., *Naseleniye Rossii za 100 let (1811–1913 gg)*, Moscow, 1956

Rubinshteyn, N. L., *Selskoye khozyaystvo Rossii vo vtoroy polovine XVIIIv*, Moscow, 1957

Rudanovskiy, Svedeniya o khlebnoy torgovle v nizovykh volzhskikh guberniyakh, *Morskoy Sbornik*, Nos. 11 and 12, 1865

Sbornik svedeniy o zheleznykh dorogakh 1867, 1869, 1871, *Dept. Zhel. Dorog-Ministerstvo Putey Soobshcheniy*

Shmeyn, A., Dvizheniye khlebnykh gruzov v yevropeyskoy Rossii po guberniyam v 1875, *Stat. Vrem.*, Ser. 2, Vyp. 16

Shubin, I., *Volga i volzhskoye sudokhodstvo*, Moscow, 1927

Statisticheskiy obzor zheleznykh dorog i vnutrennykh vodnykh putey, *Ministerstvo Putey Soobshcheniya*, St Petersburg, 1893

Statisticheskiy Sbornik Ministerstva Putey Soobshcheniya, Vyp. 1, 2, 3, 4 and 5, 1877–81

Statisticheskiy Vremennik Rossiyskoy Imperii, Series 2, Vyp. 1, 10 and 16

Svedeniya o dvizhenii po zheleznykh dorogam glavneyshikh proizvedeniy zemledelcheskom promyshlennosti v uzlovykh i konechnykh punktakh seti zheleznykh dorog v 1876, *Stat. Sborn.*, Vyp. 3

Svedeniya o vnutrennem sudokhodstve v 1859–62 in *Stat., Vrem.*, Vyp. 1

Svod statisticheskikh svedeniy po selskomu khozyaystva Rossii k kontsu XIX veka, St Petersburg, 1902

Tengoborskiy, L., *Commentaries on the productive forces of Russia*, 2 volumes, London, 1855

Trudy ekspeditsii dlya izucheniya khlebnoy torgovli i proizvoditelnosti v Rossii, 4 volumes, 1868–76

Vidy vnutrennavo sudokhodstva v Rossii v. 1837 i 1838 gg., St Petersburg 1839–9

Vil'son, I., *Obyasneniya k khozyaystvenno-statisticheskomu atlasu Rossii*, St Petersburg, 1869

Voyenno-statisticheskoye obozreniye rossiyskoy imperii, Vyp. I, St Petersburg, 1853

Voyenno-statisticheskiy sbornik, Vyp. IV, St Petersburg, 1871

Yatsunskiy, V. K., Izmeneniya v razmeshchenii naseleniya yevropeyskoy Rossii v 1727–1916 gg., *Istoriya SSSR*, No. 1, 1957

Yatsunskiy, V. K., Izmeneniya v razmeshchenii zemledeliya v yevropeyskoy Rossii c kontsa XVIII v. do pervoy mirovoy voyny, in *Voprosy istorii selskogo khozyaystva, krestyanstva i revolyutsionnovo dvizheniya v Rossii*, Moscow, 1961

Yegiazarova, I. A., *Agrarnyi krizis kontsa XIX veka*, Moscow, 1959

Yershov, G., O poseve khlebov i kartofelya v guberniyakh i uezdakh yevropeyskoy Rossii za 1870, 1871 and 1872, in *Stat. Vrem.*, Ser. 2, Vyp. 10

Zablotskiy, A., Prichiny kolebaniya tsen na khleb v Rossii, *Otechestvennye Zapiski*, Tom 52, 1847

2
Railways and Economic Development in Turkestan before 1917

Railway construction in Russian Central Asia began in 1880, at a relatively early date considering the limited economic significance of this area at the time and the overwhelmingly more urgent demands for effective railway communications within European Russia. Two hundred versts of railway were built in the newly annexed Transcaspian oblast, connecting the Turkmen village of Kizyl-Arvat with the Caspian coast, two years before an effective outlet to the Black Sea was provided for Baku oil by the Baku-Batumi line.

This paradox was essentially the result of the influence of military above economic considerations on railway construction in the borderlands of the empire. In the early 1880s Russian territory in central Asia was being clearly defined for the first time. Russian dominance in the protectorates of Bukhara and Khiva was barely a decade old and the struggle between the European powers, Britain and Russia, for influence over stagnating regimes whose territories lay between their two empires in Asia was a factor of overriding importance, with repercussions far beyond the Asiatic confines where its roots lay. At least until 1889 two considerations—the need to pacify a colonial area and to prepare for further developments in the Anglo-Russian conflict—were the dominant, if not the only significant, factors influencing the course of railway construction.

When railways in Central Asia became of economic significance their value lay in the extent to which they could promote the growing of cotton and its export to the cotton factories of European Russia and, in particular, to those of the Moscow region. It is intended to examine here three facets of this development: firstly, the stages of the growth of the railway system in Central Asia before 1917 together with an examination of what extensions were projected; secondly, the effect of the improvement of communications on the export of cotton to European Russia; and, thirdly, how far the railway system promoted the import of grain to Central Asia and thus freed irrigated land for cotton-growing.

The first railway was built in 1880–1 from Mikhaylovskiy Zaliv on the Caspian coast, south of Krasnovodsk, to Kizyl-Arvat (217 versts). This line was the brainchild of General M. N. Annenkov, whose criticisms of the Russian railway system in the 1870s had brought into being the Baranov commission.[1] The railway was to make effective communication from the Caspian coast (and hence from the Russian military base in the Caucasus) to the Turkmen strongholds in the Akhal-Tekke oasis stretching for several hundred miles in a narrow strip between the Kara-Kum desert and the Kopet-Dag mountains, as yet an undefined no-man's-land between the Russian protectorates of Khiva and Bukhara and independent Persia. The expedition of 1879 against the Turkmens had failed ignominiously for the Russian army, largely because of problems of communications. Supplies and transport could only be obtained from neighbouring Persia and of 12 273 camels used, 8377 had died on the expedition.[2] Annenkov, almost alone among the military, had supported the idea of a railway for which permission was given in June 1880. In fact, its contribution to the unexpectedly rapid success of Skobelev's campaign in the following year was limited.

The Transcaspian Railway (after 1889, the Central Asian Railway) was a military conception and remained under military control, built and exploited by the two Transcaspian Railway Battalions until 1899, when it was transferred to the

---

[1] Andreyevskiy, I, 810–11.
[2] Curzon, 37–8.

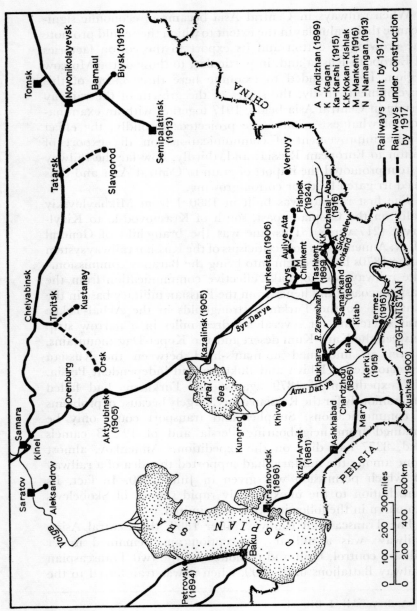

Fig. 5. Railways in Central Asia before 1917

Ministry of Communications. Its limitations, as was recognised in the report preceding its transfer to the civil authorities, largely resulted from its military origins.[1] By 1899, however, the line had acquired more than a military/political significance. In April 1885, as a serious crisis developed between England and Russia, the Tsar ordered the line to be extended from Kizyl-Arvat to the river Amu Darya, through the recently acquired lands of the Merv Turkmens, and to the borders of the settled, cultivated area of Turkestan.[2] Within a year, the 754 versts of railway line had been completed, and by the end of 1888 the line extended up the Zeravshan valley, through the dominions of the emir of Bukhara to Samarkand, a further 346 versts. These extensions gave the line that economic significance which it had lacked in its original conception.

Within a decade, isolated areas of Central Asia had been brought into relatively rapid communication with the developed parts of the Eurasian continent. The Central Asian Railway changed the whole pattern of communications in the area, not only within the Russian dominions but also within Persia and Afghanistan. It was the key factor in promoting a settled life in Transcaspia where none had existed before and hence a powerful economic influence on the previously isolated neighbouring province of Persian Khurasan. The railway was not begun with the direct intention of developing the weight of Russian influence in Khurasan, but, for all its limitations, it was bound to have the result of emphasising the isolation of that province from the ocean trade routes to the Persian Gulf.

In the 1890s the economic significance of the Central Asian Railway was further extended. Although the line had been begun in 1880 it was not until 1896 that an effective outlet to a deep-water port at Krasnovodsk was completed. Of particular commercial importance, however, was the extension of the line between 1895 and 1899 from Samarkand to Andizhan, deep in the cotton-growing Fergana valley, with branches to Tashkent (the administrative and military centre of Russian Turkestan) and to the market town of Skobelev in

---

[1] *Zapiski*, 1–9.
[2] Rodzevich, 13–14.

the Fergana valley. At the time of the transfer to the civil authorities in 1899 the strategic branch down the Murgab valley from Merv to Kushka on the Afghan frontier had just been completed. No further state-owned lines of any length were added to this 2377-verst network of the Central Asian Railway before 1917.

THE ORENBURG–TASHKENT RAILWAY

The second main line constructed before 1917 in Central Asia was the state-owned Orenburg–Tashkent line. The Central Asian Railway had served a valuable purpose in the pacification of the Transcaspian oblast and establishment of Russian authority in these borderlands. Its initial aim, however, had been to make communication effective between the Caucasus and the Transcaspian interior rather than between Central Asia and European Russia. During the 1890s in the absence of more effective direct communications the exports of Turkestan were channelled and considerably expanded through Transcaspia by the Central Asian Railway. They were then transported by sea, *either* through Baku and Batumi on the Transcaucasian Railways and again by sea to Odessa and other Black Sea ports, *or* (after 1894) to the railhead of the Vladikavkaz Railway at Petrovsk on the west shore of the Caspian and thence to the main Russian rail network, or by the Caspian and the Volga system to the central industrial areas. All these routes suffered from the extra expense incurred by transferring cargoes from one type of communication to another. Also the Central Asian Railway only reached the administrative and economic centres of Turkestan by a roundabout route. The railway did not put an end to the caravan route and as late as 1902 most of the iron and iron goods needed in Turkestan were transported, not by the Central Asian Railway, but by the more direct though slower caravan route from Orenburg to Tashkent through Kazalinsk.[1]

Thus, the Central Asian Railway did not obviate the need for communication entirely by rail between European Russia and

---

[1] *Zhurnal MPS*, 1902, No. 2, 159–60; *Sbornik materialov*, I, 68–9.

Turkestan. Rather the impetus it gave to the development of cotton-growing in Turkestan made a more direct line of communication more urgent. There were two possible routes: one following the line of the Syr Darya and passing north of the Aral Sea; and the other following the line of the Amu Darya through the Khivan and Bukharan protectorates south of the Aral Sea. The first proposed line had the limitation that for most of the distance between Orenburg and the town of Turkestan it would pass through desert and semi-desert lands. Its value would therefore be as a channel of through communication, although irrigation works from the Syr Darya might eventually make the area of greater economic significance.[1] The second line would be of more immediate value along its intermediate sections because of the cotton trade and continuing commercial importance of the two protectorates along the Amu Darya. Although the traffic was expected to be twice as heavy on this latter line as on the former, the decision was made in favour of the more northerly route on the political and military grounds that a line entirely through Russian territory would be more secure than the line through the protectorates which would be more open to attacks from hostile natives in a crisis.[2]

A railway between Orenburg and Tashkent had first been suggested in 1874. The project was approved in 1892, but not until 1900 was the route finally delineated.[3] The 1736-verst railway was begun in 1900 and opened to through traffic in January 1906. To the Orenburg–Tashkent line was added the Kinel–Orenburg line (privately built in 1877) to form the Tashkent Railway which now had its junction with the main network only 25 miles from Samara.

THE TURKESTAN–SIBERIA LINE

No further main lines were built by the State in Turkestan before 1917, but, in consonance with the policy of encouraging private investment in new railways pursued after 1905,

---

[1] *Zhurnal M.P.S.*, 1902, No. 2, 159–60.
[2] Rum, xvii–xviii.
[3] *Aziatskaya Rossiya*, II, 545, 549.

several important lines were built with private capital and others were under serious consideration. The years 1909–16 were a boom period for railway construction. Not only was the initiative of the State in examining and surveying new lines more systematic, but also private Russian capital was available to contribute substantially to financing even extensive railway projects in isolated areas.

Of the main-line projects under consideration in this period the most important for central Asia was the Turkestan–Siberia line. This project was under serious consideration by the inter-departmental New Railways Committee even before the completion of the Tashkent–Orenburg line. Four meetings in March 1905 were devoted to it, as a result of which decisions were made on the principle and surveys undertaken.[1] Deep differences of opinion on the economic and political necessity of a link between Turkestan and central Siberia were apparent in the discussions in the New Railways Committee, as among the experts of the *Obshchestvo Vostokovedeniya* who discussed the issue at several meetings in 1905–6. Nobody seems to have denied the value of linking the mineral wealth of the Altay with the Trans-Siberian line, and the value of the grain of Semirechiye to Turkestan equally seems to have been recognised. However, there were doubts about completing the link between Semipalatinsk and Vernyy.[2] Consequently, while an inter-departmental committee was formed in 1906 for a complete study of the economic potential of the whole Turkestan–Siberian line, funds were provided for full surveys only of the Novonikolayevsk–Semipalatinsk and Arys–Vernyy sections. The central section, Semipalatinsk–Vernyy, was accepted as being a project of the more distant future and consequently only very limited funds were provided for a hurried preliminary survey of it.[3] The chief argument in favour of the complete line was that the excess grain of Siberia and Semirechiye and the fuel resources of the area would enable Turkestan to specialise more effectively in cotton production and make the empire self-sufficient in cotton.[4]

---

[1] *Otchet*, iii, 1; Obshchestvo, 36–37
[2] Obshchestvo, 41–43, 85.
[3] *Otchet*, iii, v.
[4] Obshchestvo, 19–21.

In 1905–6 the Turkestan–Siberia railway was recognised as being valuable for the economic development of the area through which it passed (as would any railway, given the limitations of the Russian system), but the northern and southern sections were considered to be of greater immediate economic importance than the central section. It was thus these northern and southern sections which were promoted after 1909 when projects were put forward by private capitalists for the Altay and Semirechiye railways. The powerful consortium of Russian and French banks, formed in June 1912 to finance railway construction, was particularly interested in the Altay project from Novonikolayevsk to Barnaul and Semipalatinsk with a branch to Biysk.[1] Construction work was begun on this northern section of a Turkestan–Siberia railway on 1 June 1913 and by November 1913 the isolated section Aul to Semipalatinsk (87 versts) was temporarily opened to traffic. Just over a year later, in January 1915, the 200-verst section from Novonikolayevsk to Altayskaya was opened to temporary traffic. The 137-verst branch to Biysk was opened in the course of 1915, while the final link between Altayskaya and Aul (and thus with Semipalatinsk) was completed and in operation by May 1916.[2]

The southern section from Arys on the Tashkent Railway to Pishpek—the Semirechiye Railway—was begun in July 1912,[3] the concession being granted to the same financial group as was involved in the Altay railway. By the end of 1913 the Arys to Auliye-Ata section was under construction. By February 1916 temporary traffic had been opened as far as Mankent, 104 versts from Arys, and by August 1916 142 versts of track had been laid. In spite, however, of Kuropatkin's pressure on the War Ministry in 1916, the line was not completed to Pishpek until 1924.[4] The construction of a central section of a Turkestan–Siberia railway was only a matter of time, for, early in

---

[1] *Materialy po istorii SSSR*, VI, 583–98.
[2] *Puti*, 1915, No. 6, 654–61; *Zheleznodorozhnoye Delo*, 1916, Nos. 13–14, 123; *Vestnik*, 1914, No. 5, 45–6.
[3] *Puti*, 1912, No. 6–7.
[4] *Vestnik*, 1916, No. 6; Pyaskovskiy, 742, note 340. The Arys to Auliye-Ata section of the Semirechiye Railway appears in the official railway timetable for 1916 though no times are given.

1914, the Semipalatinsk–Vernyy line was scheduled for detailed study the following year.[1] The New Railways Committee in 1916 designated as being of prime importance (and therefore to be built in the subsequent five years) the 1400-verst Vernyy–Semipalatinsk–Slavgorod line which would have been essentially the central section of a Turkestan–Siberia link between Semirechiye and the Altay.[2]

BRANCH LINES

The Turkestan–Siberia line was the most significant line for which preparations were being made, and sections of which were being built by 1917. Other projects were completed in Turkestan in the period after 1905 as a result of the new encouragement given to private initiative by the State. Firstly, private branch lines were constructed in the Fergana valley from the Central Asian Railway to provide outlets for the cotton export in a cramped area lacking any adequate road communications. A concession was granted in 1909 for a line from Kokand to Namangan (85 versts) which was completed in 1913 to become the Fergana Railway. In the same year a concession was granted to the same company for an extension from Namangan to Dzhalal-Abad up the Naryn valley, with branches to the Central Asian Railway at Andizhan and to the market town of Kokan-Kishlak, a total of 183 versts.[3] The bonds were floated on the London Stock Exchange and the sections of the line were opened to temporary traffic during the course of 1915–16. The second important line completed in Turkestan in the new period of railway construction after 1909–10 was the Bukhara Railway. The company was founded in July 1913 by an energetic engineer, V. P. Kovalevskiy, who held a concession for a line of 585 versts running from Kagan on the Central Asian Railway (the junction for Bukhara) south-east to Karshi, with a branch to Kitab, and then along the Afghan frontier to Termez. The line was, apart from its

---

[1] *Vestnik*, 1914, No. 5, 45–46.
[2] *Zheleznodorozhnoye Delo*, 1916, No. 35–36, 238–39.
[3] *Puti*, 1909, No. 1, 9; 1913, No. 11–12, 1062–67.

strategic significance, intended to enable irrigation and cotton cultivation to be developed in southern Bukhara, and to bring Afghanistan closer by the expansion of trade relations. Construction was completed in two years, the first train reaching Termez on 9 February 1916, which was remarkable considering the difficulties of increasing costs and problems of supply during the war.[1]

Finally, the renewed interest in railway construction after 1909 saw the revival by private initiative of the idea of a link between Turkestan and European Russia through Khiva and Bukhara and the south-east corner of the Aral Sea. This project had been rejected in favour of the Orenburg–Tashkent line in 1900. A thorough economic and statistical survey was made by L. L. Rum and A. O. Brücklmeier in 1912–13 of the so-called Kaspiysko–Aralskaya railway from Aleksandrov-Gay (south-east of Saratov) to Chardzhou on the Central Asian Railway. The need for more direct access for Khivan and Bukharan cotton to the central manufacturing regions, and for the movement of grain in the opposite direction, together with the potential of the fishing industry on the Aral Sea, the oil production of the Emba region and Russian trade with Afghanistan were the basic considerations behind the project.[2]

In May 1916 the New Railways Committee under N. N. Gyatsintov considered three slightly differing projects for such a railway between Aleksandrov-Gay and the Afghan frontier via Chardzhou.[3] Among the railways of prime importance to be built within five years the committee included lines from Aleksandrov-Gay to the Emba river and from Kungrad to Chardzhou, a total of 1100 versts.[4] These suggestions encompassed most of the line proposed by Rum and Brücklmeier, and would have served the purpose of providing rail communication for the Emba oil region and for the development of Khiva and the Amu Darya valley. The produce, and in particular the cotton, of Khiva and Bukhara would still have found its outlet via the Tashkent and Central Asian Railway.

---

[1] *Byulleten*, 1916, No. 2, 65–70; *Vestnik*, No. 43, 359.
[2] Rum, xviii, 1–2.
[3] *Vestnik*, 1916, No. 24, 152.
[4] *Zheleznodorozhnoye Delo*, 1916, No. 25–26, 238–39.

ECONOMIC EFFECTS OF THE RAILWAYS

From the late 1880s when the Central Asian Railway reached Samarkand it became a powerful factor in delineating the character of the economic development of the area. The railway, as well as enabling Russian manufactures to dominate the Turkestan market, could more importantly provide an outlet for cotton which had been grown there since ancient times on irrigated land, but which previously had developed little beyond the needs of the local population.

The initial impetus to cotton growing in Turkestan was given by the conquest and pacification of the area in the 1860s and 1870s. General Kaufmann fully realised the economic value of Turkestan as a cotton-producing area for Russia. In the late 1870s and 1880s experiments were carried out to find an American, higher quality strain of cotton, suitable to the conditions of Turkestan, to replace the poor quality native strains and enable the cotton of Turkestan to compete with that imported largely from the United States.[1]

In the 1880s production of cotton in Central Asia was estimated to be 3 million poods, most of which was used locally. About 600 000 poods per annum were exported to European Russian via Kazalinsk and the Orenburg Railway. Exports were clearly on a rising trend and the figure of 831 158 poods for 1886 (i.e. before the Central Asian Railway became a significant factor) shows that exports could still increase in spite of the limitations of communications (Table 11).

The first significant point in the influence of railway communications on the cotton export and production of Central Asia was the completion of the railway to Samarkand in 1888. The obvious initial result was a sharp decline in cotton transport by caravan through Kazalinsk and thence on the Orenburg Railway although, even in 1892, 139 000 poods of Central Asian cotton still reached European Russia by this route, which was more expensive than the railway even for the still isolated Tashkent and Fergana cotton-growing regions.[2]

The year after the opening of the railway to the Amu

---

[1] Mayev, 112–16; *Trudy imperatorskogo*, 1885, I, iii, 390–91.
[2] *Sbornik materialov*, I, 32–33.

TABLE 11

TRANSPORT OF COTTON TO EUROPEAN
RUSSIA VIA KAZALINSK AND
ORENBURG

(thousand poods)

1883	603	1888	417
1884	627	1889	306
1885	667	1890	172
1886	831	1891	63
1887	545	1892	139

Sources: Rodzevich, 47–48; Mayev, 87–88; Svod zamechaniy, 160–61; Sbornik materialov, I, 68.

Darya saw a substantial increase in total cotton despatches with nearly one million poods transported from the Caspian port of Uzun-Ada.

The completion of the line to Samarkand, however, produced more striking advances, with an increase of despatches from Uzun-Ada of nearly 100 per cent in 1889 over 1888. In the following four years they more than doubled again—1·5 million poods in 1889 to 3·8 million in 1893. The largest proportion of this (1½–2 million poods) was from Turkestan proper, the rest mainly from Bukhara and Khiva, with about 100–150 000 poods per annum reaching Uzun-Ada from Persian Khurasan via Ashkhabad. This level of export was not exceeded until the railway was extended to the deeper water port of Krasnovodsk in 1896. In that year shipments of cotton from the railhead increased immediately to over 4 million poods and were maintained at a level of 4–5 million poods per annum in the subsequent three years (1896–9) until the opening of the railway to Tashkent and Andizhan. This latter was followed by a rise in the general level of cotton exports from Krasnovodsk up to 1906. In 1899 they did not exceed 5 million poods but in 1901 a record 7·7 million poods was sent to European Russia. The figures in Table 12 suggest that between 1899 and 1905 (from the completion of the

TABLE 12

Despatches of Cotton from Uzun-Ada and Krasnovodsk and Total Supply of Cotton to European Russia from Central Asia

(thousand poods)

1887	903	1,478	
1888	873	1,308	
1889	1,471	1,776	
1890	2,770	3,171	including
1891	2,667	3,001	Transcaucasian
1892	3,137	3,879	cotton
1893	3,788	—	
1894	—	—	
1895	—	—	
1896	4,164	—	
1897	—	—	
1898	4,722	—	
1899	4,907	—	
1900	4,151	5,806	
1901	7,710	6,535	
1902	6,326	6,635	
1903	4,856	5,270	
1904	6,667	6,216	
1905	5,303	6,490	
1906	7,127	8,263	
1907	5,634	9,783	
1908	4,101	7,948	
1909	4,323	10,373	
1910	4,429	10,851	
1911	6,611	11,545	
1912	6,826	11,583	
1913	4,639	12,664	
1914	6,121	14,700	
1915	—	17,500	

Sources: *Obzor* (1898–1914); *Svod zamechaniy*, 153–58; Mayev, 87; *Byulleten*, 1914, No. 4 (8), 40–41; 1915, No. 3–4, 33, 42; 1916, No. 2, 22; *Statistiki*, 2–3; *Materialy dlya izucheniya*, I, 27.

Andizhan and Tashkent branches until the opening of the Tashkent Railway) cotton was, in some years, reaching European Russia by a route other than the Central Asian Railway. This could only have been the Tashkent–Orenburg caravan route. The figures suggest that on average 400 000 poods per annum and upwards to one million poods of cotton were still reaching Russian factories by this slow and expensive route.

The opening of the Tashkent Railway in 1905 immediately increased the use of Central Asian cotton to 8·3 million poods in the first year of its operation, but nearly 7 million poods of this was still transported via the Central Asian Railway and Krasnovodsk.[1] The figure for shipments from Krasnovodsk in 1906 was second only to that for 1901. The new level of use of Central Asian cotton of roughly 8 million poods was maintained and raised substantially in the first period of the Tashkent Railway between 1906 and 1910, reaching a level of 10–11 million poods in 1909–10. This was substantially above the 5–6½ million poods level of 1900–5 before the opening of the Tashkent line.

Following the high point of 1906, cotton despatches to Krasnovodsk on the Central Asian Railway naturally showed a sharp decline to 5·5 million in 1907, levelling out to 4–4½ million in 1908–10 in spite of the trend towards rapid increase of total cotton despatches from Central Asia. But cotton traffic on the Central Asian Railway through Krasnovodsk then expanded between 1910 and 1914 reaching a level comparable with that before the opening of the Tashkent Railway. The Tashkent Railway thus only temporarily reduced the significance of the Central Asian Railway as a link between European Russia and Turkestan.

The statistics of traffic published by the Ministry of Communications did not delineate as a separate item cotton transport, giving only total traffic and figures for those goods which could be classified as 'main goods' over the system as a whole—cereals, oil and oil products, coal, salt and timber. Thus, although we can determine from the traffic figures for Krasnovodsk port the effect of the opening of the Tashkent line on the use of Krasnovodsk and hence of the main line of the Central Asian Railway as a cotton exporting route, we

---

[1] 7·13 million poods (total cotton despatched from Krasnovodsk) minus 0·25 million poods (Persian and Afghan cotton imported via the Central Asian land frontier on to the Central Asian Railway) equals 6·88 million poods.

can discover only roughly the figure for transport on the Tashkent line. By deducting exports through Krasnovodsk (minus Persian and Afghan imports over the central Asian land frontier) from total use of Central Asian cotton, we obtain the last column of Table 13. This gives approximate figures for cotton transported to European Russia by the Tashkent Railway, assuming that all that was not transported by the Central Asian Railway went by the Tashkent line.

TABLE 13

CALCULATION OF COTTON TRANSPORT ON THE TASHKENT RAILWAY
1906–14

(thousand poods)

	Total from Krasnovodsk	Imports via central Asian land frontier	Total Russian cotton from Krasnovodsk	Total central Asian cotton used	Approx. transport of cotton on Tashkent Railway
1906	7,127	253	6,874	8,263	1,389
1907	5,334	337	4,998	9,783	4,785
1908	4,101	308	3,793	7,948	4,155
1909	4,323	397	3,927	10,373	6,446
1910	4,429	478	3,951	10,851	6,900
1911	6,611	539	6,172	11,545	5,400
1912	6,826	545	6,290	11,583	5,300
1913	4,639	525	4,113	12,664	8,550
1914	6,121	510	5,611	14,700	9,110

Sources: *Obzor* (1906–14); *Statistiki*, 2–3.

The division of the cotton export between the two lines was partly a result of geographical factors. Cotton from Transcaspia, Samarkand oblast, Bukhara and also a substantial part of the Khiva harvest almost certainly went by the Central Asian Railway. However, the figures for the harvests of 1911–14 suggest that cotton, even from regions close to the Tashkent Railway, was being despatched not by that railway but via Krasnovodsk and the Volga (Table 14). Over the three years 1911–13 transport of cotton from Krasnovodsk totalled 16·6 million poods, whereas in the three seasons

TABLE 14

Sources of Cotton Traffic on Central Asian Railway via Krasnovodsk

(thousand poods)

'Russian' cotton from Krasnovodsk			Emirate of Bukhara	Production Samarkand oblast	Trans-caspia	Khanate of Khiva	Total
1911	6172						
		1911–12	1092	818	600	550	3060
1912	6281						
		1912–13	1236	806	975	522	3541
1913	4113						
		1913–14	1456	953	896	659	4137
1914	5611						

Source: *Byulleten*, 1914, No. 4 (8), 40–41.

1911–14, the total cotton harvest in the regions most likely to use the Central Asian Railway and Krasnovodsk as an outlet amounted to only 10·7 million poods. This suggests that a substantial part of the cotton harvest even from the Fergana oblast reached European Russia not by the more direct Tashkent Railway, but by the rail/river route via Krasnovodsk and the Volga system.

The construction of the Central Asian Railway from the Caspian to the cotton-producing areas of Turkestan aided an already increasing harvest to find an outlet to the market of European Russia. The limited availability of camels on the Orenburg–Tashkent caravan route would have restricted further expansion of cotton growing in Turkestan. Once the railway had reached Samarkand, however, production of cotton spiralled upwards although transport to the Moscow region was still more expensive and slower than for American cotton.[1] The chief cause of this very rapid increase must have been the impetus given to domestic cotton production (once the railway was completed to Samarkand) by the imposition of a high tariff rate on imported cotton from 1890. The extension to Andizhan and Tashkent gave a further impetus, and it was in

---

[1] *Sbornik materialov*, I, 32–33.

these areas that the most striking increases in cotton production can be seen. In Khiva and Bukhara there was no startling increase (though in general a slow upward trend) between 1893 and 1910. The increase in cotton production between 1893 and 1905 is largely an increase in production in the Fergana valley and the Tashkent area of the Syr Darya oblast. After 1905 there is also a striking increase in a new cotton-growing area—Transcaspia, where nearly one million poods of pure cotton were produced in 1914–15.[1] The development could hardly have occurred without the Central Asian Railway, constructed originally for military purposes in this remote region.

The railway system facilitated the expansion of cotton production in Russian Turkestan, Khiva and Bukhara from 6·5 million poods in 1905 to 12·7 million in the 1913–14 season. By the latter date the weakness of railway communications for cotton export to European Russia applied largely to the Khiva and Bukhara protectorates. In the latter, the Bukhara Railway promised to provide more effective access to and from southern Bukhara. A line from Chardzhou down the Amu Darya valley to Khiva, as mentioned above, was planned during the war. These lines would have promoted cotton-growing in the protectorates where development beyond a natural economy had been much slower than in Turkestan proper.

By 1914, in Fergana, where 50–55 per cent of Russian cotton was produced, communications were still inadequate for the full development of cotton growing. This inadequacy could hardly be overcome by further railway construction. With the completion of the Fergana Railway to Dzhalal-Abad and its branches, the Fergana valley would have had 645 versts of railway, giving 3·2 versts per 100 square versts, which was considered about half the norm for a territory such as Fergana. The narrowness of the valley, however, restricted the need for communications. The engineer, N. P. Petrov, suggested 9 versts of railway per 10 000 population, which would have meant 1800 versts in Fergana with a population of 2 million. But with the Central Asian and Fergana Railways encircling

---

[1] *Byulleten*, 1914, No. 4 (8), 40–41.

the valley, the only possible lines were feeders to these main lines (apart from a line from Andizhan to Osh). Gubarevich–Radobylskiy of the official Cotton Committee argued convincingly that more feeder railways would not be profitable in Fergana and that it would be more rational and cheaper to provide an effective feeder road system to the railway lines from outlying cotton-producing areas. As more and more of the irrigated lands in Fergana came under cotton (in some parts as much as 80 per cent was under cotton) so the need for an effective outlet to the main lines of communication increased, as also for inlets for almost all the needs of the population, whose economy was becoming increasingly specialised.[1]

The weakness of communication in the Fergana valley was that links between outlying cotton-growing areas and the railways broke down for two to three months between December and March, when such roads as existed were impassable. Questionnaires sent out by the Cotton Committee on the problem of communications in central Asia produced strong appeals from the Fergana valley for new roads to the outlying villages.[2]

Apart from these routes of purely local importance and improvement of communications with Kashgar, Gubarevich–Radobylskiy emphasised the need to build a good road from Pishpek over the mountains via the Kugart pass to the Naryn valley so that grain from Semirechiye could reach Fergana from two directions, over this road and along the Semirechiye Railway.[3] Gubarevich–Radobylskiy's conclusions on the need for feeder roads rather than railways only confirmed the views expressed at a series of congresses of agriculturalists and cotton growers of Turkestan between 1909 and 1914.[4]

After the completion of the main railway lines to and within Turkestan proper, the question of development of cotton production was not mainly one of providing effective outlets for cotton export. As Krivoshchein recognised after his visit to Turkestan in 1912:

---

[1] Gubarevich-Radobylskiy, 2–4, 19, 29.
[2] Ibid., 29–32, 58.
[3] Ibid., 80–81.
[4] *Materialy dlya izucheniya*, VI, 58–60.

'... at the present time this development (of cotton) is being held up by the expense and shortage of grain, particularly of fodder, the expense and shortage of labour and the expense and shortage of credit.'

Removal of these difficulties could enable cotton production to expand on the existing irrigated area. However, the long-term prospect of freeing Russia from the need to import American cotton could be achieved only by irrigating new land in the Transcaucasus and, particularly, in Turkestan.[1] In this matter developments were particularly slow in spite of the denunciation of the Russo-American commercial treaty and fears of a cotton famine.[2] As cotton growing expanded in Turkestan, the Tashkent and Central Asian Railways became increasingly important for supplying Fergana and other parts of Turkestan with grain, which it no longer produced in adequate quantities itself. However, although by 1914 Turkestan was being supplied with substantial quantities of grain from outside, the supply was erratic, expensive, liable to come

TABLE 15

EXPORT AND IMPORT OF GRAIN TO AND FROM UZUN-ADA

(thousand poods)

	Exports	Imports Grain	Flour
1888	151	—	—
1889	256	—	—
1890	266	327	753
1891	—	1014	790
1892	19	860	509
1893	—	420	471

Sources: Rodzevich, 61–62; *Materialy dlya izucheniya*, I, 27.

---

[1] *Materialy k peresmotru*, I, 18–19.

[2] See the discussions in the Cotton Committee reported in *Byulleten*, 1913, No. 2, 30–38; 1914, No. 3 (7), 21–24 and the review article, 46–47.

under the control of the money-lenders and cotton-buyers and hence unsatisfactory for the cotton-farming population.[1]

Before 1889 and in the early years of the Central Asian Railway, Turkestan was largely self-sufficient in grain and even exported small amounts from Uzun-Ada. By 1892, however, exports had dwindled and imports had substantially increased (Table 15). In Turkestan the area under cotton doubled from 52 114 desyatins in 1890 to 104 187 desyatins in 1895 and rose little above this level until 1899. Then, within three years, it doubled again to 223 785 desyatins.[2] Until 1899 grain shipments on the Central Asian Railway did not rise above the level of 1·5 million poods, which does not suggest any substantial reliance on imported grain, the total needs of the area being at least 40 million poods.[3] How far did increased exports of cotton between 1898 and 1905 from 4·7 million poods to a high of 7·7 million in 1902 result in increased traffic in grain from the outside and from one station to another of the Central Asian Railway?

Grain despatches from Krasnovodsk towards Andizhan rose sharply from 658 000 poods in 1901 to 5·3 million in 1903, falling back to 2·2 million in 1905, while total grain transport on the Central Asian Railway rose from 1·5 million poods in 1899 to 7·0 million in 1905. The key year appears to be 1901, the year of a record cotton export of 7·7 million poods. This was not accompanied by any substantial imports of grain through Krasnovodsk (only 658 000 poods),[4] but there was a sharp increase in despatches of grain from the Zeravshan valley, now able to reach Fergana more effectively by railway. However, the following years (1902–4) saw a sharp decline in the production of cotton at the same time as a startling increase in the import of grain through Krasnovodsk, reaching

---

[1] *Byulleten*, 1916, No. 2, 18–36.
[2] Ibid., 1914, No. 4 (8).
[3] *Materialy dlya izucheniya* I, 25. Figures for grain despatches and arrivals on the Tashkent and Central Asian Railways are taken or calculated from the figures in *Stat. sborn. M.P.S.*, vyp. 64, 72, 80, 84, 88, 92, 101, 105, 112, 128, 132, 138 (for 1899–1913).
[4] This is to some extent an artificial figure as the demand for grain was present at this time but the breakdown of the railway bridge over the Tedzhen river stopped traffic on the Central Asian Railway and almost brought famine to Turkestan. *Byulleten*, 1916, No. 2, 34.

a peak of 5·3 million poods in 1903. The differences between despatches of grain from the Zeravshan valley and arrivals in the Fergana valley show that a substantial proportion of the grain imported via Krasnovodsk was reaching as far as the Fergana valley (Table 16). Grain imports to that area through

TABLE 16

GRAIN DESPATCHES ON CENTRAL ASIAN RAILWAY AND
ARRIVALS IN FERGANA VALLEY

(thousand poods)

	Desp. from Krasnovodsk	Desp. from Zeravshan valley	Arr. Kokand	Arr. Andizhan	Total arr. Fergana
1899	—	—	203	125	447
1901	658	2,562	2,063	384	3,194
1902	2,470	2,233	1,615	330	2,735
1903	5,330	1,132	2,219	311	3,492
1904	4,645	2,069	2,649	651	4,781
1905	2,218	2,730	1,996	657	4,041
1906	1,951	1,515	2,767	842	4,397
1907	1,290	3,916	3,093	1,284	6,867
1908	605	6,236	2,577	929	6,718
1909	482	6,702	3,158	1,953	9,874
1910	1,792	2,095	4,367	4,560	14,979
1911	3,305	2,889	4,467	4,088	15,044
1912	3,761	5,218	4,205	3,824	14,597

Source: Stat. sborn. M.P.S.

Krasnovodsk fluctuated between 1900 and 1905 not in relation to the production of cotton (the 'high' of 1901 was accompanied by low grain imports), *but in relation to the success of the grain harvest in the Zeravshan valley*. Thus, 1903 saw a low level of cotton export (4·8 million poods), a low level of grain despatches from the Zeravshan valley (1·1 million poods) accompanied by a record level of grain arrivals in the Fergana valley of 3·5 million poods. Whereas the level of cotton export fluctuated in succeeding years, as did the imports of grain via Krasnovodsk, the imports of grain by railway to the Fergana valley did not fluctuate but rather increased regularly from 0·4 million poods in 1899 to 4·8 million in 1904.

The grain and cotton traffic adjusted only slowly to the opening of the Tashkent Railway in 1906. Expectations that it would supply Fergana with grain and enable more cotton to be grown were not initially realised and this partly explains the pressure for the construction of the Semirechiye Railway. At first sight grain shipments on the Tashkent Railway in its early years seemed to confirm the expectations. Thus in 1905 and 1906 they were 21·2 and 23·6 million poods respectively. However, the figure for 1905 shows that 21·2 million poods were transported *before* the through connection with Tashkent was opened. Similarly, the largest part of the grain traffic on the completed Kinel–Tashkent line was from the section between Orenburg and Kinel, down the valley of the Samara river from which grain and flour were exported largely *not* to Turkestan but to European Russia and beyond.[1] This is shown by the limited arrivals of grain at Tashkent and the junction with the Central Asian Railway before 1910 (Table 17).

TABLE 17

GRAIN TRAFFIC ON TASHKENT RAILWAY, 1905–13

(thousand poods)

	Arr. Kinel from Tashkent direction	Desp. in both directions			Arrivals		Desp.
		Orenburg	Sorochinskaya	Buzuluk	Tashkent town	Tashkent Rly. junct.	Arys
1905	18,837	4,984	5,170	4,256	—	—	—
1906	14,344	4,886	4,214	3,967	175	1,426	35
1907	6,776	1,505	1,859	2,189	206	1,437	733
1908	14,740	2,903	2,560	3,134	199	407	1,511
1909	37,319	7,966	7,294	7,833	266	2,625	1,568
1910	23,246	7,529	8,625	10,269	2,451	16,657	588
1911	9,998	4,549	4,181	5,173	2,049	8,773	367
1912	16,502	4,579	4,596	6,377	1,649	8,000	596
1913	40,129	—	—	—	—	—	1,781

Sources: *Stat. Sborn. M.P.S.*; *Vestnik*, 1914, No. 30, 269.

[1] *Vestnik*, 1914, No. 30, 269.

The failure of grain to penetrate to Turkestan in the first years of the Tashkent Railway from the Samara valley and Orenburg area was compounded by fluctuations in the imports of grain through Krasnovodsk. These dropped between 1905 and 1909 from 2·2 million poods to 0·5 million poods. On the other hand, 1907 to 1909 were years of record despatches of grain from the Zeravshan valley (3·9 million poods in 1907 to 6·7 million in 1909).

The problem of organising a transport system which would enable cheap grain to reach Turkestan and thus free irrigated land for cotton was one which continued to be discussed in these first years of the Tashkent Railway. The Turkestan–Siberia link was discussed by the New Railways Committee and surveys were undertaken. The *Obshchestvo Vostokovedeniya* devoted several meetings to the topic and the newly established official Cotton Committee discussed it at length in 1907–8, although the weakness and vagueness of the statistical material available made any clear evaluation of the problem difficult.

In the *Obshchestvo Vostokovedeniya* before an audience which included senior officials of the Ministries of Finance and Foreign Affairs and of the *Glavnoye Upravleniye Zemleustroystva i Zemledeliya*, L. M. Goldmerstein outlined the economic significance of the Turkestan–Siberia line for Turkestan. In Semirechiye alone, he argued, there was a surplus of about 18 million poods per annum of grain. Prices in Vernyy were only 25–30 kopecks per pood whereas in Turkestan they had risen from 45 kopecks per pood in 1890 to 1 ruble 35 kopeks in 1901. Only difficulties of transport from Semirechiye, 'the natural granary of Turkestan', hampered cotton production in Turkestan. The Syr Darya and Fergana oblasts needed about 16 million poods of imported grain to maintain the existing level of land under cotton. Goldmerstein also pointed to the wastefulness of exporting cotton to European Russia and cotton goods back to China and Persia for export. The Turkestan–Siberia railway would provide Turkestan with cheap fuel to enable it to develop its own weaving and spinning industries:

'The transport of coal and cheap grain should provide the possibility of developing on a large-scale textile production in Turkestan itself.'

More daringly he suggested that the line could lead to the export of grain via the Central Asian Railway, Baku and Batumi to southern Europe.

Some of his audience, at least, were not convinced of the value of the through connection. N. P. Fedorov doubted whether the 250 million rubles would be well spent and suggested rather branches to the Trans-Siberian Railway and the development of irrigation in Turkestan for cotton growing. Was there any evidence, he asked, that the population of Turkestan would or could reduce the area under grain and increase that under cotton if cheap grain was available from the north? Fedorov was not alone in doubting the value of planning the entire Turkestan–Siberia line at a time of State bankruptcy, and while the country was faced with the problems of political, military and naval reform.[1]

At the third meeting of the Cotton Committee (4 May 1907) the issue was raised again at the request of seventeen

TABLE 18

TRANSPORT OF GRAIN TO TURKESTAN AND FERGANA VIA TASHKENT RAILWAY

(thousand poods)

	Despatches of grain: Kinel towards Tashkent	Despatches of all goods towards Tashkent			Central Asian Railway Despatch of grain	
		Orenburg	Soroch-inskaya	Buzuluk	From Tashkent town	From Tashkent Rly. junction
1906	3,364	2,821	193	415	2,445	859
1907	3,944	2,161	395	488	1,121	1,227
1908	373	—	—	—	559	558
1909	159	2,854	349	335	1,188	2,513
1910	4,297	6,824	4,812	3,125	747	12,482
1911	11,098	5,710	2,024	2,098	647	12,966
1912	10,389	5,253	896	2,309	752	8,106
1913	7,108	643	906	—	—	—

Source: Stat. Sborn. M.P.S.

[1] Obshchestvo, 8–43.

Moscow cotton manufacturers, proposing a line from Omsk through Semipalatinsk and Vernyy to Tashkent. Their main argument was that the area under grain in Turkestan would not be reduced under present conditions as the native population had, in some years, taken risks to get the high profit on cotton only to find a bad harvest in Russia prevented grain transport. Only a Turkestan–Siberia railway would give the security of supply necessary. However, N. P. Fedorov again opposed the through link, pointing out that even if all Turkestan had to be supplied with grain it would need only 45–50 million poods, which would not justify the expense.[1]

Tables 17 and 18 show that the Tashkent Railway in its early years, 1906–9, was not a channel for the supply of Turkestan with substantial deliveries of grain. In particular, the grain markets of Sorochinskaya and Buzuluk despatched very little towards Tashkent. There was a sharp change in the situation in 1910, which was continued at least for the next three years. From 1910 substantial imports of grain clearly came from the Orenburg–Kinel area to Tashkent and were transported further to the Central Asian Railway and the Fergana valley, as the figures for 1910–13 in Tables 16–18 show. The rise in grain imports from 1910 was accompanied by a rise in the area under cotton, particularly in Fergana, where there was an increase from 187 733 desyatins in 1908 to 291 752 desyatins in 1914. The same years brought a rise in the general level of use of Central Asian cotton by Russian factories from 7·9 to 14·7 million poods, about 55 per cent of total consumption, although the volume of consumption of imported cotton declined only as a result of the outbreak of war.[2]

Figures for the increase in sowings of cotton in Central Asia suggest that increased grain imports from 1910 to 1913 were improving the situation, but that the supply of grain was only one of the factors involved. The figures in Table 19 show virtually no increase in sowings in Samarkand which was also a grain-producing area but which, because of its height above sea level, was less favourable for cotton than Fergana and parts of the Syr Darya oblast. The general level of sowings

---

[1] *Trudy*, I, 27–35.
[2] *Byulleten*, 1916, No. 1, 15.

TABLE 19
Increase of Sowings of Cotton in Turkestan
(desyatins)

	Fergana	Samarkand	Syr Darya	Transcaspia	Bukhara and Khiva
1909	208,053	31,269	31,463	26,035	80,000
1910	235,891	25,224	35,675	28,328	110,000
1911	267,347	28,666	67,838	32,882	82,000
1912	266,566	27,024	64,150	42,319	87.000
1913	274,897	31,758	76,726	46,512	105,000
1914	291,752	32,363	73,915	44,739	92,000(?)

Source: Byulleten, 1914, No. 4 (8), 59–60; 1916, No. 2, 19.

of cotton in Bukhara and Khiva remained within the 80–100 000 desyatin bracket, although there were fluctuations. In Transcaspia there was a clear and substantial increase in sowings, but this area was not affected by grain supplies down the Tashkent Railway. The substantial increases in sowings appeared in the Fergana and Syr Darya oblasts where they nearly doubled in 1911 and maintained this level, more or less, in succeeding years. These oblasts were the very ones which suffered from lack of grain.

In spite of these favourable results there was dissatisfaction at the highest level with the achievements. Krivoshchein visited Turkestan in 1912 and in his report concluded that one of the factors holding up cotton production was still the shortage and expense of grain. Discussions in the Cotton Committee also reflected this dissatisfaction. Kh. N. Sergeyev, a high official of the *Glavnoye Upravleniye Zemleustroystva i Zemledeliya*, asked what could be done in view of the fact that the Tashkent Railway, although important, had not justified the great hopes put on it. Grain was not being supplied to Russian Turkestan from Siberia and European Russia. Zagorskiy, representing the Ministry of Finance, agreed that there were limitations on the Tashkent Railway. Tariffs did not encourage internal transport of grain over more than 540 versts and thus it was more profitable for the grain producers of the Orenburg and lower Volga areas to sell in European Russia. Certainly by this time, as we have seen, the Tashkent Railway was slowly stabilising its traffic with increased transport of grain along the line and a two-way movement of cotton to the north and grain and timber to the south was making the railway more profitable. However, expensive grain transport on the Tashkent Railway was seen only as a partial answer to the problem. The New Railways Committee, in approving the proposals for the building of the Semirechiye Railway, expected extremely large grain shipments from that area to Turkestan, which might even threaten the Tashkent Railway with a cut in traffic. Zagorskiy thought that from the point of view of the increase of cotton growing the Semirechiye Railway would fully solve the problem.[1]

The insecurity of the situation was shown by the fact that although there had certainly been a sharp increase in 1910 in the areas under cotton and in production, which coincided with the first large grain transports along the Tashkent Railway, and although the high level of grain imports was maintained in the following year (1911), prices were so high that they led to some decrease in the area under cotton in 1912. This was particularly true of the Syr Darya and Fergana oblasts, where cotton cultivation was most intense.[2]

The limitations of the statistical material for the war years prevent us from examining in detail the further developments in the grain/cotton situation. However, it is clear that there was a substantial increase in domestic cotton production from 13 million poods in the 1912–13 season to approximately 18·85 million poods in 1915–16.[3] Clearly there were particularly favourable circumstances—the difficulties of importing cotton from the United States and high prices on the international market favoured high domestic production. However, it is more difficult to say how far in this situation there was a solution of the grain problem. The local cotton farmer, because of his indebtedness to the cotton buyers, found it difficult to turn back to producing grain even if necessary, as he often bound himself to produce cotton to be sold entirely to the cotton buyer in return for credit or grain at artificially high

---

[1] *Byulleten*, 1913, No. 2, 30–38.
[2] Ibid., No. 2, 15–16.
[3] Ibid., 1916, No. 3, 30–31.

prices in the sowing season.[1] If the war period resulted in a reduction of grain imports (and the figures for 1913 suggest that even in this year large grain despatches to Turkestan along the Tashkent Railway were not maintained),[2] with the increase in the production of cotton which is evident there may have been a serious grain-supply problem in Turkestan. The building of the Semirechiye Railway was too slow to ease the situation during the war.

The existence of the railway system, although encouraging the growing of cotton and thus the development of the economy of Turkestan, may, on the other hand, because of its inadequacies, have led to a deterioration in the position of the native cotton grower in a period when the pressures to grow cotton in the main areas were irresistible. The cotton grower could not escape from the vicious circle of pressure to grow cotton and artificially high grain prices. The import of grain by the railways, therefore, did not of itself encourage the native population to extend the area under cotton. It did not remove the danger of starvation when the Turkestan grain harvest was small; and in 1916 threatened to lead to a massive abandonment of cotton growing.[3]

---

[1] *Byulleten*, 1916, No. 2, 25.
[2] *Vestnik*, 1914, No. 30, 269.
[3] *Byulleten*, 1916, No. 2, 18–36.

BIBLIOGRAPHY

Andreyevskiy, I. E. (ed.), *Entsiklopedicheskiy slovar'*, 43 volumes, St Petersburg, 1914

Aziatskaya Rossiya (Pereselencheskoye Upravleniye, Glavnoye Upravleniye Zemleustroystva i Zemledeliya), 3 volumes, St Petersburg, 1914

Byulleten Khlopkovogo Komiteta (*Byulleten tsentral'nogo Khlopkovogo Komiteta*—1915–16), St Petersburg, 1913–16

Curzon, G. N., *Russia in Central Asia in 1889 and the Anglo-Russian question*, London, 1889

Gubarevich-Radobyl'skiy, A., *Kakiye dorogi sleduyet stroit' v Turkestane i Zakavkaz'e* (Materialy dlya izucheniya khlopkovodstva, IV), St Petersburg, 1914

Materialy dlya izucheniya khlopkovodstva (Glavnoye Upravleniye Zemleustroystva i Zemledeliya: Khlopkovyy komitet), Vyp. I–VII, St Petersburg, 1912–17

Materialy k peresmotru russko-amerikanskogo torgovogo dogovora, Vyp. I–III, St Petersburg, 1912

Materialy po istorii SSSR, VI ('Dokumenty po istorii monopolisticheskogo kapitalizma v Rossii'), ed. A. L. Sidorov *et alia*, Moscow, 1959

Mayev, N., *Turkestanskaya Vystavka 1890 g.*, Tashkent, 1890

Obshchestvo Vostokovedeniya, Sredne-Aziatskiy Otdel, *Protokoly tryokh zasedaniy, Turkestano-Sibirskaya zheleznaya doroga*, ed. P. A. Rittikh, St Petersburg, 1906

Obzor vneshney torgovli Rossii po evropeyskom i aziatskom granitsam (Department tamozhennykh sborov), St Petersburg, 1802–1917

Otchet o rekognitsirovannykh izyskaniyakh zhelezno-dorozhnoy linii Semipalatinsk-Vernyy (Iliyskiy pos.) protyazheniyem 980.79 v. proizvedennykh osenyu 1907 g. ekspeditsiyeyu inzhenera Glezera (Ministerstva Putey Soobshcheniya: Upravleniye po sooruzheniyu zheleznykh dorog), St Petersburg, 1908

Puti soobschcheniya Rossii (Zhurnal otdel statistiki i kartografii Ministerstva Putei Soobshcheniya), St Petersburg, 1908–17

Pyaskovskiy, A. V. (*et al.*), *Vosstaniye 1916 goda v Sredney Azii i Kazakhstane, sbornik dokumentov*, Moscow, 1960

Rodzevich, A. I., *Ocherk postroiki zakaspiyskoy voyennoy zheleznoy dorogi*, St Petersburg, 1891

Rum, L. L., Brücklmeier, A. O., *Kaspiysko-Aral'skaya doroga v ekonomicheskom otnoshenii*, St Petersburg, 1914

Sbornik materialov sobrannykh sredne-aziatskogo zheleznodorozhnogo ekspeditsiyeyu, vyp. I, St Petersburg, 1894

Statisticheskiy sbornik Ministerstva Putey Soobshcheniya, St Petersburg, 1877–1917

Statistiki bumagopryadil'nogo i tkatskogo proizvodstva za 1900–10 gg. (Ministerstvo Torgovli i Promyshlennosti, Otdel Torgovli), St Petersburg, 1911

Svod zamechaniy komanduyushchego voyskami zakaspiyskoy oblasti po revizii zakaspiyskoy voyennoy zheleznoy dorogi v 1894 godu, St Petersburg, 1895

Trudy imperatorskogo vol'nogo ekonomicheskogo obshchestva, St Petersburg, 1885

Trudy khlopkovago komiteta (G.U.Z. i Z. Department Zemledeliya), I–II, St Petersburg, 1907–8

Vestnik putey soobshcheniya, St Petersburg, 1900–16

Zapiski o sostoyanii zakaspiyskoy zheleznoy dorogi i glavneyshikh potrebnostyakh eyo uluchsheniya, St Petersburg, 1899

Zhurnal Ministerstva Putey Soobshcheniya, St Petersburg, 1900–5

Zheleznodorozhnoye Delo, St Petersburg, 1900–17

3

The Soviet Concept of a Unified Transport System and the Contemporary Role of the Railways

Recent Soviet textbooks on economic and transport geography have devoted considerable space to the discussion of a 'unified transport system', usually prefixed by a brief statement of the difference between the transport systems of socialist and capitalist countries. Lavrishchev, expressing a typical Soviet view, states that

'Under the capitalist system of economy the irrational distribution of production accounts for the irrational distribution of the means of transport. The sharp division of the capitalist countries into industrial and agrarian, economically highly developed and backward, the uneven distribution of production within individual capitalist countries, and the isolation of industrial and agricultural production from the areas of consumption lead to irrational transportation of raw materials and fuel, half-finished and finished products, and in the end result in enormous waste of public wealth and reduce the productivity of social labour. In many capitalist countries the transport is owned by large monopolies. The capitalist transport is characterised by competition among the different private transport companies—railways, shipping lines, bus and truck lines, airlines—which redoubles the irrational work of the transport in the capitalist world';

whereas, he argues,

'In socialist society the transport is the common property of the people and is a constituent part of the single socialist system of economy. The systematic, proportional development of socialist economy conditions the rational development and distribution of all forms of transport over the territory of the country. The distribution of the transport in its turn fosters a systematic distribution of production all over the country. . . . Unlike the elemental and anarchic development of the means of transportation in the capitalist countries, the railways, waterways, automobile and air transport have developed in the U.S.S.R. as a single system of transportation which systematically combines all forms of transport and works in accordance with a plan established by the state. One of the most important differences between the socialist transport of the U.S.S.R. and the transport of the capitalist countries is that the former develops without any crises and its freight turnover is continuously increasing.'

Reference to the concept is also found in Saushkin's *Ekonomicheskaya geografiya SSSR* (Moscow 1967) and at somewhat greater length in Kazanskiy's *Geografiya putey soobshcheniya* (Moscow 1969) among other works. Similar views are expressed in a Polish work by Tarski called *Koordynacja transportu* (Warsaw 1968). Such a view of transport as a unitary whole is a contrast to the discussions in earlier Soviet textbooks where almost the entire work was devoted to railways and only the most superficial regard was given to other media, quite without reference to the interdependence of the media (e.g. Sarantsev's *Geografiya putey soobshcheniya* (Moscow 1957)). A similar pattern was followed by textbooks of economic geography, including the longstanding *Ekonomicheskaya geografiya SSSR* by N. N. Baranskiy (Moscow, several editions). The basis of the concept is to regard transport as a unitary whole within the framework of the production cycle of the economy. It produces the effort to move commodities between the different stages of production and ultimate consumption. Because transport is not in a material sense 'productive', it is essential to reduce the effort required to handle these flows to a minimum in order to reduce the input of human and material resources to the lowest possible level within the transport infrastructure so that such resources may be used in more materially productive sectors of the economy. To achieve such a 'minimum input—maximum output' equation, the mechanism of central planning is used to make sure that the commodity flows and passenger movements in volume and direction are handled as rationally as possible by the transport media most suited to a given parameter. Soviet authors, in extolling the virtues of the concept, point out that such a system has all the wasteful competition eliminated but may leave room for constructive inter-media competition. The media are not considered as separate entities in the system but as an interrelated, interdependent and interdigitated structure, so that for any flow the 'media mix' is governed by factors such as distance, volume, frequency and orientation. Kazanskiy and several other authors suggest that movements are classifiable as 'local', 'inter-regional', 'trunk' and 'international', while a distinction is made between the highly localised 'industrial' haulage and the general pattern of national movements classed as 'general haulage'. Consideration must also be given to the nature of the terminal facilities necessary for different media when the selection of the mix for any given flow is made. There is, thus, an awareness of *scale* in terms of volume and distance in the transport effort.

Transport provision, however, has been distinctly affected by policy decisions made in the late 1920s, when basic Soviet planning concepts were being formulated, and implemented by *Gosplan* in the inter-war and immediate post-war period. It is only in the most recent times that these decisions have been modified, but, no doubt, accelerated change will take place in the last quarter of this century. A search for a high degree of self-sufficiency has produced powerful introvert trends in the transport system that have orientated it towards interior domestic producers and consumers rather than to the ports and border crossings for participation in the broader pattern of world trade and transport. Consequently the Soviet transport system is 'continental', with a marked dependence on land-based media, while such 'continentality' produces a distortion in operating distances compared to those commonly regarded as optimal for land-based media, reflected in the remarkably long average hauls by railway and waterway.

An early decision was taken to disperse industry as evenly as possible among the regions and in the third Five-Year Plan the first serious attempts were made to develop a full industrial infrastructure in each of the major planning regions. This was thought strategically desirable but raised critical transport issues, especially in the more remote and underdeveloped regions of Siberia, the Far East and Central Asia. One of the attractions of the dispersal of industry and regional self-sufficiency was the theoretical economy in transport effort by moving industry towards its raw materials and fuel bases, so that reduction of unnecessary hauls and elimination of wasteful crosshauls became a key principle in regional design. Unfortunately, the uneven distribution of fuel and mineral resources make regional self-sufficiency hard to attain, while the more complex the economy becomes and the greater the degree of product diversification, the less easily achievable such an aim proves to be. It was found in practice that a strict limit was placed on the degree to which industrial dispersal could be carried, partly for production-planning reasons and partly because of the transport implications. Too high a degree of dispersion would have created a massive and insuperable demand for transport routes and facilities. Because of the problems of the organisation of industry and transport, dispersal took the form of major industrial nodes so that large-scale industrial complexes were developed in a limited number of areas with a reasonable selection of the requisite raw materials and locational factors. The effect was to concentrate transport demand on a limited number of trunk routes, giving economies of scale for both industry and transport as well as obviating the need in many cases to build new routes. The great 'metallurgical bases' that typified the gigantomania of the 1930s and the more recent idea of the 'territorial production complex' may be seen as the product of this concept of 'nodes'. The early abortive but grandiose plan for the Ural-Kuzbass *Kombinat* had transport as the key to its realisation, though it failed because of inadequate investigation of resources available in the Ural and in the Kuzbass. The discovery that iron ore was available near to the Kuzbass coalfield made Ural supplies less necessary, while the development of railways to the Karaganda coalfield gave the Ural works a source of fuel nearer than the Kuzbass, though it was not qualitatively so good. The distribution of the major industrial nodes fits well into the T-shaped pattern of arterial routes.

The selection of the primacy of industry rather than of agriculture brought not only a need to reorientate the tsarist transport system developed in a strongly agricultural setting but also conditioned the selection of transport media for development, particularly by the concentration on heavy industry. The sophistication of the Soviet economy and its arrival at the threshold of the 'post-industrial society' are likely to change such emphasis between now and the end of the century. So far, the priority accorded to heavy bulk freight traffic—coal, coke, ore and petroleum or timber—imparted a relatively simple commodity flow of a narrow range of items needed in large quantities and delivered in a form of 'moving belt technique' over average hauls of several hundred kilometres. These bulk goods still comprise 85 per cent of all shipments by tonnage and 80 per cent of the goods-traffic turnover of the four main media, while the pattern of flow arising from the distribution of producers and consumers gives a strong east–west emphasis. Natural waterways by their predominantly north–south alignment have been unable to satisfy demand for primarily east–west shipments, so that the railways have been the mainstay of this transport system.

The nature of any transport system is affected by the dimensional framework within which it operates and it is the vastness of distance and area that distinguishes the Soviet system. With a total area of 22·4 million km² the Soviet Union straddles northern Eurasia. It lies across a distance of 4500 km from north to south, from Arctic desert through tundra and tayga to steppe and desert. From its westernmost border in the north European plain to the eastern outposts on the Pacific Ocean, it stretches across 9000 km of continental terrain. It is a country with vast, sparsely settled lands, for over two-thirds of the Soviet population live in the one-sixth of the country's area south of Leningrad and west of the Volga, into which over half of the total national investment is poured; although the U.S.S.R. is ninety times greater in area than the United Kingdom, its population is little more than four times greater. From any distribution map of demographic or economic phenomena, a clearly marked predominant distribution within a triangular area with its base between the Baltic and the Black Sea and its apex in western Siberia can be seen: it is within this *oecumene* between the colder boreal forest to the north and the poorer steppe to the south that the overwhelming demand for transport takes place (Fig. 6). Beyond this *oecumene* only a few scattered patches of well developed and modestly peopled country lie in the 'empty quarters' of Siberia and Central Asia.

For its latitude, because of its continental situation, the Soviet Union is anomalously cold—long cold winters, broken by strong blizzards, are typical, while in the far north-east some of the world's lowest temperatures are recorded. The southwestern interior is distinguished by evaporation greatly exceeding precipitation with a consequent tendency to aridity, but winters are cold despite the heat of the summers. In contrast, with evaporation lower than even the meagre precipitation and natural drainage impeded by factors of the terrain, the north is generally excessively wet. Forty-seven per cent of the country's area is liable to the occurrence of permafrost phenomena that make civil engineering works in the north and east problematical. While it is perhaps imprudent to press geographical determinism too far, there is doubtless a case in the Soviet Union that climatic factors exert considerable influence over the choice and use of transport media.

If transport flows are plotted on a linear basis of direct links between origin and destination, the inter-regional pattern of commodity flows shows a distinct lattice-like structure (Fig. 7). In real terms, such movements generally are fed into the main arterial system through which they move for some part of their journey before diverging to their destination, as reflected by diagrams for individual commodities published in several Soviet textbooks but based on Nikolskiy's work in the early 1960s (Fig. 8). The main arterial system is

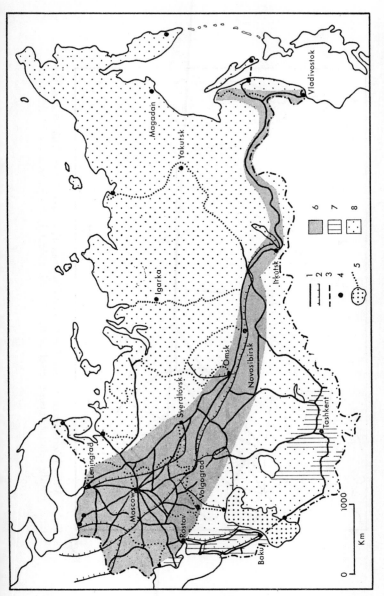

Fig. 6. Kazanskiy's concept of the unified transport system and belts of major economic activity. 1. Arterial railway routes. 2. Trunk pipelines. 3. Railway ferries. 4. Major ports. 5. Main waterways. 6. Territory with dense population and high concentration of productive investment. 7. Subsidiary territory of the same characteristics. 8. Sparsely settled and weakly developed territory.

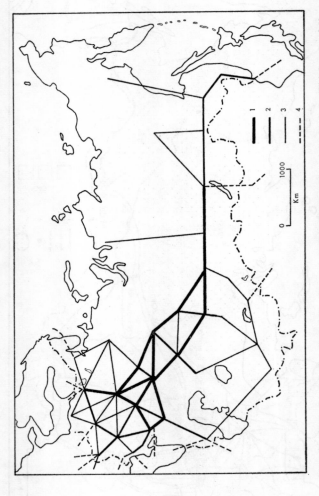

Fig. 7. Main interregional freight flows (modified from Kazanskiy). 1–3. Thickness of line represents the relative importance of the link. 4. Main freight flows with areas outside the Soviet Union.

T-shaped, in which the head of the T may be seen as the north–south movement axis along the railways between the north-western industrial area, the Central Industrial and Central Blackearth regions and the Industrial South, augmented by flows along the Volga and Dnieper waterway systems. A further southerly extension may be considered to be traffic moving into the Caucasian lands by rail or by water across the Black Sea and Caspian Sea. From this markedly 'European' concentration, the stem of the T may be distinguished in the strong arterial movements along the east–west axis from the Central Industrial region on the west, across the Volga lands and the Urals to western Siberia and Baykalia, continuing on a much diminished scale eastwards to the Far East. The T-shaped figure fits neatly within the 'major economic belt' defined by Kazanskiy (Fig. 6), which corresponds closely with the triangular *oecumene* already described. All the evidence of these flows suggests that about half the area of the country is not involved to any significant extent in the transport system; in fact, 55 per cent of the country lies more than 100 km from a railway line.

THE TRANSPORT BALANCE

The railways provide almost two-thirds of the traffic effort for freight and just about half the traffic effort for passengers; while road haulage is responsible for only about a twentieth of all freight though contributing about a third of the passenger traffic (Table 20).[1] The effect of the short hauls by road transport on traffic effort is reflected in the very large proportion of originating tonnage and passengers handled by road transport compared to the railways. Waterways contribute modest shares of freight and passenger traffic, while a rising volume of traffic has been handled by pipeline in a closely defined range of commodities. The generally long hauls of freight by sea-going ships are shown by the exceptionally modest share of originating goods but an appreciable share of total freight traffic effort: sea-going passenger traffic is

---

[1] See also tables and discussions in concluding chapter.

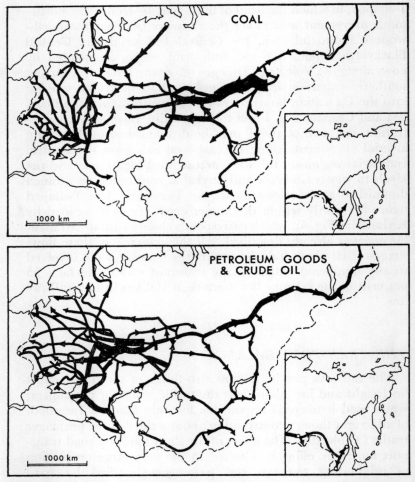

Fig. 8. Freight flows of selected commodities by railway (based on Nikolskiy).

very small, mostly restricted to the Black Sea, Caspian Sea and Far Eastern coastal shipping.

Airways hold a marked share of passenger traffic and a rapidly growing number of originating passengers, but in freight haulage their share is exceptionally small and most air freight represents mail or newspaper printing plates set

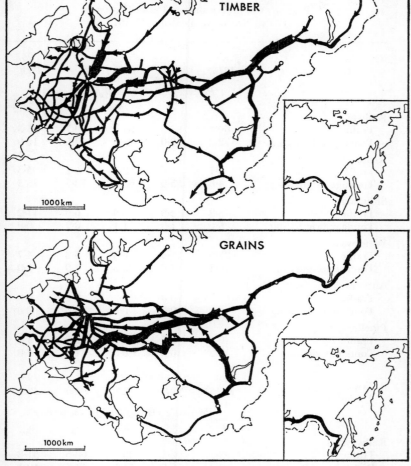

Thickness of line is proportional to volume of freight. Data for early 1960s.

in Moscow but printed in provincial centres. Most striking are the short hauls by road transport and by sea-going ships for passengers (Table 21). These represent movements in towns and local hauls in the countryside for road transport, while ferries and short pleasure trips are the principal elements at sea. The very long overseas movements of cargo by sea-going

TABLE 20

PERCENTAGE SHARE OF THE PRIME HAULIERS
IN TOTAL NATIONAL TRANSPORT, 1913–67

Goods transport	1913*	1940	1950	1960	1967 Plan
	Percentage share of total for the country				
	%	%	%	%	%
Railways					
Traffic	60·4	85·2	84·4	79·7	66·3
Tonnage originating	72·3	38·0	29·5	17·5	16·9
Roads					
Traffic	0·1	1·8	2·7	5·2	5·3
Tonnage originating	4·6	55·0	65·6	78·7	78·3
Shipping					
Traffic	16·3	4·9	5·5	6·9	17·2
Tonnage originating	6·8	1·9	1·1	0·7	0·9
Waterways					
Traffic	22·9	7·4	6·5	5·3	4·7
Tonnage originating	16·1	4·6	3·3	1·9	1·9
Airways					
Traffic	—	0·0†	0·2	0·2	0·6
Tonnage originating	—	0·0†	0·0†	0·0†	0·0†
Pipelines					
Traffic	0·1	0·7	0·7	2·7	5·9
Tonnage originating	0·2	0·5	0·5	1·2	2·0

Passenger transport	1913*	1940	1950	1960	1967 Plan
	Percentage share of total for the country				
	%	%	%	%	%
Railways					
Traffic	93·2	92·4	89·5	68·5	51·8
Passengers originating	94·4	66·5	51·1	14·5	10·1
Roads					
Traffic	0·0†	3·2	5·4	24·6	34·8
Passengers originating	0·0†	29·5	46·4	84·3	89·0
Shipping					
Traffic	3·0	0·8	1·2	0·5	0·4
Passengers originating	1·4	0·4	0·3	0·1	0·1
Waterways					
Traffic	4·0	3·5	2·7	1·7	1·2
Passengers originating	4·2	3·6	2·2	0·8	0·6

CONCEPT OF A UNIFIED TRANSPORT SYSTEM

TABLE 20 (*continued*)

Passenger transport	1913*	1940	1950	1960	1967 Plan
	Percentage share of total for the country				
	%	%	%	%	%
Airways					
Traffic	—	0·1	1·2	4·7	11·8
Passengers originating	—	0·0†	0·0†	0·1	0·2

* In contemporary boundaries.   † Share too small to allocate.
Source: *Transport i svyaz SSSR*, Statistika, Moscow, 1967.

ships account for an average haul over 4000 km. In both rail and waterways movements goods hauls are much longer than for passengers, though the movement of commuters tends to reduce the railway passenger's average journey and the same is done for waterway transport by pleasure boats. It is perhaps worth noting that over 65 per cent of all goods sent by rail move over 200 km while distances over 500 km account for 44·0 per cent of all goods. Although it is expressed policy to transfer all hauls of under 80 km by railway to road transport, 20 per cent of all goods sent by rail still move less than 100 km. It is, on the other hand, now recognised

TABLE 21

AVERAGE LENGTH OF HAUL BY PRINCIPAL MEDIA

	1969	
	Goods (*Km*)	Passengers (*Km*)
Railway	861	91 (27)*
Road	17	7·5
Sea	4047	48
Waterways	486	37
Airways	1310†	1000†

* Commuting traffic
† 1968
Sources: *Narodnoye khozyaystvo SSSR v 1970 godu*, Moscow, 1971. *Ekonomicheskiy spravochnik zheleznodorozhnika*, Moscow, 1971.

that road haulage up to 200 km is to be reckoned with. The predominance of the railways in the statistics viewed collectively is evident: the Soviet Union claims to handle 45 per cent of the world's railway freight turnover. The average railway freight traffic density in the Soviet Union is an annual 16 million ton. kms/kilometre of track, five times the average for the United States. A leading American railway man described the density of freight traffic on the West Siberian section of the Trans-Siberian Railway as 'unbeatable on American railroads'. It is also claimed that the number of passengers per kilometre of track is several times the world average. Considering the continental nature of the Soviet Union, such a situation is not unexpected, particularly when it is remembered that the rivers in general flow at right angles to main demand orientation.

Part of the continuing dominance of the railways arises from a clear policy decision of the 1920s not to encourage road transport—a decision conditioned by the attraction of economies in investment and resources that would accrue from developing an existing medium rather than encouraging growth of an embryonic medium—while the strength of the railway lobby perpetuated this situation long after the reality of the initial decision had faded. Yet whatever struggles for power are reflected in the railways' present strength there is the undeniable fact that the railway, more than any other medium, suits the environmental conditions and the demand for transport in the U.S.S.R. Roads and waterways are still regarded primarily as feeders to the railways; pipelines and airways are viewed as a means of relieving the railways' task in certain critical areas of freight or passenger demand.

RAILWAY CONSTRUCTION AND OPERATION

Few relief obstacles appear to stand in the way of easy railway construction and operation, especially as the Ural mountains have numerous easy passageways through their central section, in southern Siberia comparatively easy routes exist through the rough, dissected country of Baykalia, and the littoral shelf provides a way to circumvent the mountainous terrain of the Great Caucasus. Three-quarters of the total

CONCEPT OF A UNIFIED TRANSPORT SYSTEM

route is straight track and has gradients gentler than 1:166. Only one-fiftieth of the route has gradients steeper than 1:60. It is the micro-features of relief that present problems of civil engineering: the gullying of the steppe; the swamps of the broad and ill-drained river basins; the drifting sands of the deserts; and even the great rivers to be bridged. Bridges over 60 m in length comprise 2 per cent of the total number but 22 per cent of the total length of bridges on Soviet railways. It is unfortunate that original economies in building costs have left routes that now hamper operations by imposing severe speed and weight restrictions. But, in the last analysis, it is perhaps distance that is the major limitation to new building: one kilometre of single track requires 100–170 tons of steel rail, 185 m^3 of wood sleepers and 1500 m^3 of ballast, scarce in the great plains. Even a branch line may demand a massive resource input when it is as long as the 720 km route from Tayshet to the Lena river—equivalent to London to Perth.

Railway operations are markedly affected by climatic conditions. Most serious is the great winter cold that causes increased fuel consumption because of intense heat loss; raises the viscosity of diesel fuel, so making special fuel supply systems necessary; freezes wet bulk loads (ores, etc.) or points and equipment; and even causes metal fatigue. Aridity in the south and the winter cold in the north long posed a water supply problem for steam locomotives. An advantage of diesel and electric traction is the 90 per cent saving in water consumption. Open plains provide no obstacles to strong winds—consequently there is a need for high-cost shelter belts so that in 1950 there were 37 450 km of shelter belts along railways, with a total area of 101 200 ha. High winds and freezing present difficulties with overhead equipment on electrified sections, quite apart from the common problem of freezing of points. Upset of the thermal balance of the ground by civil engineering structures in areas liable to permafrost can produce detrimental warping and slumping. Almost everywhere, special attention has to be given to the roadbed drainage because of distortion in the winter freeze, a task made more difficult by the scarcity of good ballast. Throughout the country, although the railways are carried on low embankments above flood level in the spring thaw, the distortion of the track after the winter freeze demands a period of speed restriction until the track is corrected. The swollen rivers with their ice floes in spring require bridges of adequate length and strength to withstand the strain. In some places where bridges are too costly for the traffic to be sustained or because of other site problems, ferries are used (for example across the Amur at Komsomolsk; across the Straits of Kerch; and across the Caspian Sea). Nevertheless, railways appear to be less likely to be interrupted by these conditions than other media; rivers become so swollen and beset by ice floes that navigation must cease, as it does also in the low water of late summer; road transport is made uncertain by snow and ice, especially by snow drifting in blizzards; while the largely unpaved roads are quagmires in the spring thaw and dust ribbons in summer.

The commitment to industrialisation after the Revolution demanded an increased role to be played by the railways and a consequent change in the operational character of the railway system. In contrast to the pre-revolutionary pattern of short, light and infrequent trains serving a predominatly agricultural economy, heavier, faster trains with a greater service frequency were needed. Under the influence of GOELRO,[1] railway electrification was discussed at an early stage, but generating capacity was slow to develop and first priority was given to 'productive' industry. Consequently, apart from the electrification of suburban railways around Moscow, the heavily graded Suram Pass route (1932) and the Kizel–Sverdlovsk line in the Ural mountains (1932–9), dependence was put on the steam locomotive. Experiments were conducted with diesel traction in Central Asia where there was an acute water shortage for steam locomotives, but there was little follow-up and the condensing tender began to be used. In the prevailing conditions the steam locomotive was attractive because it was cheap and easy to build, simple to maintain and operationally reliable, while there was no shortage of fuel.

The growth of traffic was in some respects the downfall of the steam locomotive. Increasing frequency of trains demanded costly double-tracking and the provision of more

---

[1] State Commission for the Electrification of Russia, formed 1920.

passing places and, even if the increase in frequency were to be offset by heavier trains, the use of more powerful steam locomotives would demand massive strengthening of the track. This would have been an insuperable task in time and money over the distances involved on main and secondary routes in the Soviet Union. Alternatives to the big increase in axle load needed with rigid-frame locomotives proved unsuccessful; the monstrous AA-20 class 4–14–4 locomotive did immense damage to track and points, while articulated locomotives proved too costly to maintain, despite their relatively low axle loads.

Both diesel and electric traction gave far better power-weight ratios so that heavier trains could be handled faster without appreciable increase in axle loads. Even though diesel and electric locomotives and their trackside installations were more expensive than steam traction, compared to the cost of massive track strengthening for heavier steam locomotives they both gave worthwhile savings in time and resources. Growing output of electric current in the late 1950s made electric traction more desirable, while French experience with 25 kv showed a way to save valuable copper by lighter catenary and to reduce the number of trackside substations. Soviet sources stress the saving in copper, one of the scarcer metals in the Soviet bloc. Electrification has been pushed ahead on trunk routes where traffic density has warranted the high cost, so that the length of electric traction has risen from 3000 km in 1950 to 33 900 in 1970. The pattern shows a striking similarity to the T-shaped arterial system already discussed. Diesel traction, with its lower trackside costs, has been widely used since the improvement in the Soviet supply of petroleum and increase in refinery capacity in the 1950s, while it has owed much to the application of American experience. In 1950 diesel traction operated 3100 km of route compared to 76 200 km in 1970.

Distance intensifies the operating problems of Soviet railways compared to more compact systems. There are exceptionally long hauls for many commodities, though the overall figure—in 1970, 861 km—is kept low by the considerable proportion of freight moved over short distances that would go by road vehicles in most western countries (Table 22).

TABLE 22

DISTRIBUTION OF RAIL FREIGHT BY DISTANCE MOVED

km	per cent of total freight
up to 100	20·5
101– 200	13·6
201– 500	21·7
501–1000	16·0
1001–2000	16·4
2001–3000	6·8
over 3000	5·0

Source: *Ekonomicheskiy spravochnik zheleznodorozhnika*, Moscow, 1971.

Goods such as petroleum, iron and steel, wood and grain or mineral fertilisers have average hauls of over 1000 km. The average haul of coal is 692 km, five times longer than the corresponding movement in the United Kingdom. At the other extreme, baled cotton moves 3327 km and salt fish 2271 km. On such long journeys wagons are many days on the way and empty running becomes a serious problem: the annual loss of capacity through empty running amounts to 130 million tons of goods. Refrigerated wagons do 47 per cent of their running empty, open wagons about 26 per cent, while the average for all types is 38 per cent. Elimination of empty running is all the more important with high-capacity wagons which have enabled a 25 per cent increase in train weight to be achieved without longer trains or greater axle loads, avoiding heavier track or longer sidings. In terms of passenger traffic, to run one train a day each way on the Trans-Siberian Railway requires at least eighteen train sets, where as a comparable London–Aberdeen service could manage with two sets.

Like nearly all railway systems the 135 200 km of railway in the Soviet Union show a very uneven intensity of use. Soviet sources quote figures such as 46 per cent of the route length carries 86 per cent of all freight traffic, while 64 per cent of all freight traffic moves on the quarter of the route length which is double track. No comparable statements are made for passenger traffic but it is clear from the Soviet railway timetable that, away from the trunk routes, density of passenger trains falls to three or less each way daily. This may be compared with the *Beeching Report* which noted that 65 per cent of all passenger traffic moved on 13 per cent of the route length of British Railways, while 8 per cent of the route length took 38 per cent of all freight traffic. Other evidence of concentration on Soviet railways is illustrated by statements such as 80 per cent of all empty wagons are loaded daily by six railway directorates—Donets, Tomsk, Pechora, Karaganda, Sverdlovsk and Kirov; all but two are coal-mining areas, reinforcing the fact that 60 per cent of all wagons loaded daily are for coal transport. Eight per cent of Soviet goods stations despatch 80 per cent of all coal, while 55 per cent of all consignments of iron ore come from the Krivoy Rog district on the Dnieper railway—eleven stations despatch 70 per cent of the ore, while 72 per cent is received by only fourteen stations, reflecting a highly concentrated pattern of movement.

RECENT RAILWAY DEVELOPMENT

Among the more advanced countries the Soviet Union is remarkable in having a railway system that still shows vigorous growth. The overall length of route grew by almost 30 000 km between 1940 and 1970, while in the 1971–5 Plan a further addition of nearly 5000 km of track is expected. There have, however, been regional contrasts in growth, though the nature of the Soviet statistics precludes a detailed analysis of these differences. It should be recalled that statistics of route length refer to railways controlled by the Ministry of Communications (in 1970, 135 000 km) and do not include 'industrial railways', which had a total length of 120 800 km and a route length of 85 700 km in 1970. In some republics small changes in length of route may reflect changes in ownership of these lines.

In Belorussia there has been a small increase in route since 1950, but the length still remains below the 1940 figure because lightly used routes or lines which have lost their strategic significance have not been rebuilt, while some lines—notably narrow-gauge track—have been closed. In Estonia the system grew slightly compared to pre-war, but since 1960 lines have been abandoned and the total length is now less than in 1940 by 30 km. In several republics the route is longer than in 1940, but the increase has been small and often took place in the early post-war years: in Georgia there has been no increase since 1965; in Lithuania railways began to contract after 1960 but the rationalisation seems to have ended by 1965. In Moldavia a brief and minor contraction about 1960 has been followed by modest but continuing growth. In Kirgizia the length of railways has not changed since 1950, while the same has happened in Tadzhikistan. There has been no new building in Armenia and in the Turkmen Republic since 1969.

The main increases in length has come in the R.S.F.S.R. which contains over half the total route and where almost two-thirds of the increase in length has taken place. A formidable growth has been shown by Kazakhstan whose railways have almost doubled in length since 1940, the result of new lines in the Karaganda mining area, in the Virgin Lands and on the eastern shore of the Caspian Sea. Even the Ukraine has had a 10 per cent increase in length since 1940, though territorial expansion accounts for most of this and there has been some elimination of surplus route, again mostly narrow gauge.

If a detailed historical comparison of railways in the major economic regions of the R.S.F.S.R. could be made, it would show little or no increase in European Russia apart from the Volga lands, Northern Caucasia, the Ural region and the North-east. Several short lines in the area around Lake Ilmen, as well as some along the Belorussian border, appear to have been abandoned. The overwhelming increase in new track has been in Siberia and most of the major projects yet to be completed remain in the 'eastern regions'. Important branch lines are planned to penetrate the new oilfields of the west Siberian lowlands, where construction over the wet ground poses many problems and the large rivers demand massive bridges: an example of these lines is the 700 km Tyumen–Surgut railway. Several branch lines are being built from the Trans-Siberian Railway north into the tayga of the Angara-

Ilim basins to exploit mineral wealth. Large hydro-electric projects like the Bratsk scheme and the Yenisey barrage at Maklakovo have been dependent on the construction of long branch lines.

The most important railway project is the so-called North Siberian Railway running through the tayga north of and roughly parallel to the Trans-Siberian Railway and crossing the Patom and Aldan country to the Pacific Ocean. It appears to be a resurrection of a project in the third Five-Year Plan (1938) for a Baykal–Amur Magistral along a similar alignment, of which the Tayshet–Lena and Komsomolsk–Sovetskaya Gavan sections were actually built. The present proposal follows the same general alignment, but some versions appear to extend westwards to a terminus at Tavda and to a Pacific terminus at the mouth of the Amur. It was fear of Japan in Manchuria and the vulnerability of the Trans-Siberian Railway that prompted the original line in 1938, so that the continuing strategic weakness of that railway in relation to a militant China may underlie the revival of the project.

Other proposed routes, besides branch lines to open new resources, are certain regional links which are still required. These include lines from the lower Volga to the Ural region and to northern Kazakhstan as well as completion of the long-standing project to join the Saratov area to the Amu Darya basin at Kungrad via the Emba oilfield and the Ustyurt Plateau. While the missing link of the Circum-Caspian Railway between Astrakhan and the Emba oilfield is still scheduled for completion, the proposed link from the Pechora coalfield to the Ural region appears to have been shelved.

Any study of Soviet railways' construction projects is bedevilled by 'traveller's tales'. These include a line from Salekhard on the lower Ob to Norilsk; from Volgograd via Stavropol across the Great Caucasus to Tbilisi; and a fantastic project for a line along the Okhotsk coast northeastwards to Anadyr. Another unconfirmed project is to extend the proposed line from Bolshoy Never to the Chulman coalfield north to Yakutsk and to Tiksi at the mouth of the Lena. The magnitude of these projects and the views expressed in discussion of the development of northern Siberia in Soviet journals makes their realisation highly unlikely.

Sheer extension of route length represents only one facet of growth. Plans for extensive double-tracking and electrification are perhaps equally good indicators, while technological improvement in faster trains or turbine propulsion, containerisation and computer control demonstrate continued reliance on railway transport, which is expected to handle twice its present volume of traffic by A.D. 2000.

INLAND WATERWAYS

Rivers were traditional lines of movement in the Russian lands, serving important trade flows in foodstuffs from surplus to deficit regions. As they flow either northwards or southwards they are unable to serve effectively the modern predominance of movement on an east–west axis. Rivers constitute a total length of 3 million kilometres, yet only 142 000 km are used for shipping (of which 74 000 km have a guaranteed minimum depth). Artificially improved waterways and canals measure 17 500 km and handle 60 per cent of all waterborne freight traffic. Over half of all freight turnover is on the Volga system, serving the vital north–south 'European' axis. Of the 322 million tons moved on waterways in 1968, 53·5 million tons were interchanged between waterways and railways. Apart from not having the right orientation for modern demands, the waterways fail as a 'moving belt' because of a frozen period of four months or more, except in the far south; navigation is subsequently hindered for weeks by the spring thaw, while operations are commonly halted in late summer by low water. In Soviet economic conditions it is difficult to build up a stockpile for the closed season and consequently, despite official encouragement to use waterways, a preference for railway transport is strong. Consequently, there has been duplication of river routes by the building of parallel railways (for example, along the Volga). Interlinking of river systems by canals might increase the usefulness and attraction of water transport, but such canals are extremely costly and justified only where heavy traffic or strategic considerations are present (for example Volga–Don, Volga–Baltic canals). Improvement schemes have not always been successful: the

Volga barrages give a guaranteed depth of channel, but lengthen the frozen season by ten days through slowing the flow of water, while waves form in strong winds to make the older, shallow-draught river vessels difficult to handle and siltation in large shallow reservoirs is a hazard. An 'advantage' of river transport listed in Soviet textbooks is that the shipping for a given volume of goods requires ten times less metal than an equivalent railway. There are plans for a 'unified European waterway system' embracing the Volga, Dnieper, Danube, Elbe, Oder, Vistula and other rivers that suggest that waterways are still regarded as highly significant in the Socialist-bloc transport system.

ROAD TRANSPORT

Soviet textbooks emphasise the function of road transport as short-haul distribution in both urban and country areas and as feeder services to railway yards and river wharves. It is generally viewed as a haulier operating up to 80 km, but some recent Soviet textbooks discuss its advantages in 'door-to door' transport up to 200 km and inter-city bus services are now operated over comparable distances. It is worth noting that the average haul of goods in kilometres by road is fifty times less than in rail transport; inter-city buses are still predominantly on routes shorter than 120 km. The Soviet view is that road transport can only be expected to grow as motor vehicle manufacture grows, but it has a serious disadvantage in demanding a higher labour input than other media, which is important in regions short of labour. According to Kazanskiy, one-third of all road traffic takes place in the Central Industrial and Central Blackearth regions, in the Industrial South and in the Ural region, while three-quarters of total freight traffic and two-thirds of total passenger traffic is concentrated in European Russia, with the large towns particularly significant. It is also important in petroleum-producing regions and in northern forest areas. In the 'eastern regions', as indicated in Soviet atlases, roads are regarded in many cases as extensions to the railway system and are designated as 'routes without rails'. Urban bus traffic shows a concentration in European Russia but inter-city buses are particularly important in southern Siberia and central Asia, though they are most intensely developed in areas with good road systems—the Baltic states, the western Ukraine and in Caucasia.

By western European standards, the comparative length of Soviet 'motor roads' at three times the length of railways is low. In the United Kingdom the length of roads exceeds railway route length nineteenfold. Of 1·4 million kilometres of 'motor road' in 1967, only 431 000 km were hard surfaced. Grit or dirt roads turn to quagmires in winter and spring or to dust ribbons in the summer dryness, while 'metalled' roads suffer acutely from frost. In Siberia's permafrost areas the *corduroy* road is common—a log structure covered by grit and suited to lorries. The hard frozen ground of winter favours road transport and in Siberia 'winter roads' appear across otherwise impassable country and even frozen rivers are used by vehicles. At the same time, intense cold presents problems of fuel viscosity, heat loss from engines, hardening of rubber tyres and the difficulty of keeping the cooling water warm! It is possible to talk of a road network only within the *oecumene* and in the more thickly settled parts of Central Asia and Caucasia, while simple dendritic systems mark most of Siberia. No major road yet exists right across the country from east to west: in fact, available evidence suggests that it is impossible to motor right across Siberia. In eastern Baykalia and in Amuria only local road systems focused on the Trans-Siberian Railway exist, though they often extend many hundreds of kilometres northwards into the tayga (for example, the Aldan Highway). It is important to remember that the pattern of road transport does not necessarily arise from failing technological standards but from a policy adopted thirty to forty years ago that avoided investment not only in the motor-vehicle industry, but also in road construction. Though various plans have been put forward to cure the country of its 'roadlessness', none has progressed far and targets have seldom been achieved. As a comparable situation, it is worth noting that the limitations placed by distance on successful road haulage are reflected in the state of the transcontinental railways which is healthier than the state of those east of the Mississippi in the United States, while in the United Kingdom it has been shown that a modern freightliner is a successful competitor with road lorries on distances exceeding 100 miles despite the favoured position which road haulage enjoys in this country.

AIR AND SEA TRANSPORT

The speed of air travel makes overall cost by *Aeroflot* from Moscow to eastern Siberia less than by soft-class train and saves a whole man-week in time, while it is in Arctic and desert regions that aircraft have become prime movers. Soviet observers consider that development of mining and industrial projects in remote parts of Siberia might be more cheaply served solely by aircraft (or by freight-carrying airships) than by provision of conventional land surface media, especially when the problems of terrain to be overcome in road and railway construction are considered and the climatic limitations on operations, in spring and winter in particular, are taken into account. Aircraft form, expectedly, a long-distance haulier (average haul over 1000 km for passengers and 1310 km for freight). Over 60 per cent of all air freight moves in the railless north and east. It is claimed that the equipment of an air route (including airfields, navigational aids, etc.) is only 10 per cent of the cost of a well-built road or railway. Although the Soviet Union was undoubtedly 'air-minded' in the 1930s, limitations through technological failings restricted realisation. The rise in air transport may be closely correlated in time with the growth of the Soviet aircraft industry, a product of wartime and post-war experience, aided by the use of German and American technology respectively, captured in the closing stages of the Second World War or received through wartime aid.

Sea transport has very long freight hauls (over 4000 km) and 90 per cent of all goods sent by sea move to or from places outside the Soviet Union. Domestic movements are overwhelmingly in the same sea, on the Caspian Sea, Black Sea or in Pacific coastal waters (Table 23).

TABLE 23

SHARE OF GOODS TURNOVER BY SEA AREAS IN 1960

Sea Area	Internal Movements		External Movements	
	Tonnage %	Traffic %	Tonnage %	Traffic %
Black Sea—Sea of Azov	38·9	35·6	57·0	69·8
Baltic Sea	5·3	4·8	23·3	15·1
Northern seas	4·2	7·1	9·8	6·3
Far Eastern seas	15·6	28·9	9·3	8·7
Caspian Sea	36·0	23·6	0·6	0·1
TOTAL	100·0	100·0	100·0	100·0

Internal movements = movements between Soviet ports.
External movements = movements between Soviet and foreign ports.
Source: *Ekonomicheskaya geografiya SSSR*, eds. Nikitin, N. P., Prozorov, E. D., Tutykhin, B. A., Moscow 1966.

THE UNIFIED SYSTEM

Figure 9 is an attempt to construct a model of the 'unified transport system' according to the part played in it by the major long-distance haulage media and the feeder hauliers in the interstitial space. Most clearly developed is the railway reticule in European Russia with long projecting 'feelers' into the less developed territories of the European and Siberian North, Central Asia and the Far East, though as noted, it serves less than half the country's area. Over much of Siberia rivers form the main discrete lines of transport, with an important series of intersections with the railway in southern Siberia, while lorry roads form continuations of the rivers and railways into the interior. Roads perform the same function beyond the railways in Central Asia, as well as forming valuable links across the Great Caucasus. In European Russia we may distinguish the discrete lines of transport provided by rivers augmenting railways, notably in the Volga and Dnieper basins. In a few sea areas—the Caspian, the Black Sea, the

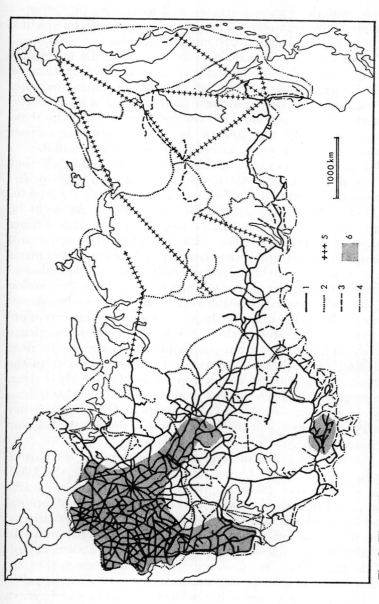

Fig. 9. The unified transport system—an essay. 1. Railways. 2. Waterways. 3. Major 'routes without rails' in less developed areas. 4. Main sea routes. 5. Important interregional air links beyond the main system. 6. Interstitial areas of well-developed road haulage as a feeder to other media.

Baltic and the Pacific fringe as well as the Arctic Ocean—shipping provides an extension of the system, often forming a link between isolated local sub-systems and the main national system. In most areas roads may be seen to provide the interstitial transport service. While airways provide a special long-distance function *superimposed* over the surface transport system, they may be seen to be an *extension* of the unified system as a whole in the remote areas of northern Siberia and Central Asia. In these areas air freight becomes an important part of transport operations. If adequate statistical material were available to allow a weighting of the individual vectors, it would show that within the 'All-Union' system there were the two major orientations that produced the T-shaped system already referred to.

How stable does this structure appear to be? As might be expected, there is a reasonably stable situation with distinct long-term trends in keeping with technological progress and change in the nature of the economy and consequent demand for transport. There has been the steady erosion of the railways' share of traffic, but even more marked has been the erosion of the traffic moved by waterway. Road transport has shown a steady if modest growth; the great 'motor vehicle explosion' has not yet been seen in the U.S.S.R. If American experience is relevant, this important threshold may, however, be near at hand, especially as the recently expanded capacity of the motor-vehicle industry comes into full production. It is clear, however, that the Soviets seek to avoid the wasteful 'American style' motor-car stage, forcefully criticised by Khrushchev during his journeys in the United States, though the solution is not yet clear.

Fluctuations in the share of sea transport reflect the participation of the Soviet Union in world trade. Since the 1960s the airways have taken a growing share of passenger traffic but future development will depend on the growth of internal and international long-distance movements by Soviet citizens, and the expansion of the sort of sophisticated cargoes now sent by air freight in the western world. Though the statistical basis of the traffic share of pipelines seems a little suspect, they are undoubtedly of growing importance as the network is extended.

While the Soviet authorities seem prepared to give forecasts for the distribution of total traffic between media, they seem reluctant to make prognostications on the composition of future freight traffic. It is unlikely that coal and coke will remain such major items as the use of natural gas and petroleum grows, though certainly 'mineral building-materials' are liable to remain a large item for a long time (transport is reckoned to comprise 30–40 per cent of their cost). It is likely that any 'consumer boom', even if more modest than in the western world, could greatly increase the movement of high-value manufactured goods whatever the intentions of Soviet regional planners. Changes in the transport system may also be caused by new directions in economic thought in *Comecon*: this could certainly happen if Comecon were to enter more fully into the world pattern of trade and accept normal and sophisticated ways of trading, which some members—like the Czechs—are anxious to do. A more cost-conscious Comecon drawn into dependence on the general level of world commodity prices would force change on the Soviet Union.

Containerisation and the bulk carrier, which have already begun a fundamental adjustment in western concepts of industrial location and commodity flow, might well force a basic rethinking of the classical planning precepts in the U.S.S.R. This would involve a final collapse of the autarkic principles that led to development of high-cost interior producers and a swing to the increasing importance of media-breaks at port sites (though few bulk-carrier ports are yet available), while there might well appear a major world 'price shed' somewhere in Baykalia. It could make sense to supply eastern Siberia and the Far East with seaborne Canadian wheat instead of railborne Ukrainian grain; European Russia could be drawn into closer relations with the rest of Europe and the Atlantic trading sphere, while Siberia might become a part of the Pacific trading sphere with a large Japanese participation. On the other hand, the increasing interlinkage economically between Europe and the east Asian trading sphere means that the Trans-Siberian Railway offers a short, fast rail route for bulk containerisation. The value of the

---

[1] Some 40 000 return passages over this route were reported as expected in 1974 by a Swiss forwarding agent and further expansion should follow the completion of a new marine terminal on the Pacific coast of the U.S.S.R. in 1975.

great-circle air routes across Soviet territory needs no emphasis. Clearly, the implications for the unified transport system as at present being expounded are considerable as the Soviet economy changes from its classical form as a virtually closed system to a part of a wider world pattern.

BIBLIOGRAPHY

Atlas *avtomobil'nikh dorog SSSR* (various dates), Moscow
Atlas *shkem zheleznikh dorog SSSR* (various dates), Moscow
Baranskiy, N. N., *Ekonomicheskaya geografiya SSSR*, Moscow, 1955
Blackman, J. H., *Transport development and locomotive technology in the Soviet Union*, Columbia, 1957
Ekonomicheskiye svyazi i transport, Voprosy Geografii, 61, Moscow, 1963
Galitskiy, M. I., et al., *Ekonomicheskaya geografiya transporta*, Moscow, 1965
Garbutt, R., *Russian railways*, London, 1950
Gumpel, W., *Das Verkehrswesen Osteuropas*, Cologne, 1967
Hunter, H., *Soviet transportation policy*, Cambridge (Mass.), 1957
Hunter, H., *The passenger automobile in the Soviet economy*, Columbia, 1964
Hunter, H., *Soviet transport experience*, Washington, 1968
Kalinin, V. K. et al., *Obshchiy kurs zheleznikh dorog*, Moscow, 1970
Kazanskiy, N. N. et al., *Geografiya putey soobshcheniya*, Moscow, 1969
Khachaturov, T. S., The organisation and development of transport in the U.S.S.R., *International Affairs*, 21, London, 1945
Khachaturov, T. S., *Ökonomik des Transportwesens*, Berlin, 1962 (trans. from Russian)
Khanukov, E. D., *Transport i razmeshcheniye proizvodstva*, Moscow, 1956
Kim, M. P., et al., *Zheleznodorozhniy transport v gody industrialisatsii SSSR 1926–1941*, Moscow, 1970
Lavrishchev, A., *Economic geography of the U.S.S.R.*, Moscow, 1968
Mellor, R. E. H., Motive power and its problems on Soviet railways, *Locomotive*, 65, 1959
Mellor, R. E. H., Some influences of physical environment on transport problems in the Soviet Union, *Advancement of Science*, 20, 1964
Mellor, R. E. H., *Die Sowjetunion—VI Das Transportwesen*, Munich, 1971
Minsker, S. S. et al., *Voprosy razvitiya zheleznodorozhnogo transporta*, Moscow, 1957
Nikitin, N. P. et al., *Ekonomicheskaya geografiya SSSR*, Moscow, 1966
Nikol'skiy, I. V., *Geografiya transporta SSSR*, Moscow, 1960
Rakov, A., *Lokomotivy zheleznikh dorog Sovetskogo Soyuz*, Moscow, 1955
Sanilov, S. K., *Ekonomika transporta*, Moscow, 1956
Sarantsev, P. L., *Geografiya putey soobshcheniya*, Moscow, 1957
Saushkin, Yu. G., *Ekonomicheskaya geografiya SSSR*, Moscow, 1967
Shafirkin, B. I. et al., *Ekonomicheskiy spravochnik zheleznodorozhnika*, Moscow, 1971
Slezak, J. O., *Breite Spur und Weite Strecken*, Berlin, 1963

Spravochnik Passazhira, Moscow, 1968
Tarski, I., *Koordynacja transportu*, Warsaw, 1968
Ukazatel' Zheleznodorozhnykh Passazhirskikh Soobshchenii—annually, Moscow
Westwood, J. N., *History of Russian railways*, London, 1964
Westwood, J. N., *Soviet railways today*, London, 1963
Williams, E. W., *Freight transportation in the Soviet Union*, Princeton (N.J.), 1962
Zvonkov, Y. V., *Principles of integrated transport development in the Soviet Union*, Chicago, 1957

4

The Soviet Merchant Marine

The vast empire of tsarist Russia, which occupied the strategic heartland of the Old World, was often identified as 'an economic appendage to Europe', for peninsular Europe drew even more heavily on the continental plains of Russia for essential imports of grain than she did on the rich prairies of the New World. In the five years 1909–13, for example, Russia provided nearly a third of the world's wheat exports, nearly three-quarters of the barley exports and nearly half of the oats exports as well as a substantial volume of timber, flax and beet sugar.[1] In 1913 Russian ports handled 41 million tons of cargo—70 per cent of it in foreign trade, and of this, 70 per cent was made up of exports, the commodity unbalance being dictated by the need to import capital for industrial development and for the redemption of foreign loans. By 1897 the grain port of Odessa had become Russia's third city, outstripping Kiev in population.

The bulk of this massive commodity outflow—one of the largest in its contemporary world—was carried in British ships; indeed, the fortunes of many British tramp shipping companies were founded on the Black Sea and Baltic trades. In 1913 only 8 per cent of foreign exports and 14 per cent of foreign imports were carried in Russian ships.[2] Well over half of the Russian coasting vessels were sailing ships, dispersed among the Black Sea, Baltic, White Sea and Far

---

[1] Stamp, 1933, 98–107.
[2] Alexandersson and Norstrom, 1963, 141.

Eastern seaboards. Deep-sea merchant shipping in those days was an indicator of an economically developed nation, though it is strange that Count Witte's investment in Russian railways was so liberal and his appreciation of the importance of good rail connections to the ports was so acute while at the same time the tsarist government's promotion of ports and shipping was on such a modest scale.

War and civil war further attenuated the resources of the merchant fleet. Rudoi and Lazarenko[1] estimated that 'in 1918, 79 per cent of merchant transport tonnage consisted of sailing and sailing-motor vessels which carried less than 1 per cent of the world's tonnage'. The Soviets' task of reconstruction began almost from scratch as far as the merchant marine was concerned. In the 1920s progress was extremely slow. Of course, external trade was then on a limited scale, priority was understandably given to railways and canals in the deployment of the 10 per cent of the work force that was then assigned to transport and the ambitious plans for heavy industry placed minimal demand upon sea transport. Deep-sea shipbuilding was not resumed until 1924 and the serious programme got under way only during the second Five-Year Plan (1933–8). The essential continentality of the U.S.S.R., the fragmented nature of the coastal periphery, the physical problems of Arctic navigation and the awkwardness of the southern sea route to the Soviet Far East forced maritime affairs to a low priority. So, also, did the Soviet policy of economic nationalism and the external economic barriers that non-recognition of the Soviet regime by many countries engendered. The nadir was reached in the early 1930s when these handicaps were reinforced by world depression. Harbron states[2] that by 1931 only 4 per cent of exports were carried in Russian hulls, and although the proportion was to increase slowly through the 1930s, the total value of exports shrank from 4539 million rubles in 1930 to 1832 million in 1934 and 1332 million in 1938.[3] Ironically, it was during this period that a more purposeful period of maritime growth began and, no doubt, there was a threefold stimulus to this process; the

---

[1] Rudoi and Lazarenko, 8.
[2] Harbron, 1962, 140.
[3] Schwarz, 1954, 590.

crucial need for a more favourable balance of payments, to be achieved in part by using more Russian vessels for imports; the example of Germany, Italy and Japan, who were aggressively expanding their oversea shipping services; and the ability to buy second-hand ships abroad at rock-bottom prices during (and after) the Depression. Russian shipyards, too, were getting into a modest stride with the series production of cargo and fishing vessels. On these bases the gross tonnage of the Russian fleet in 1939 must have been in excess of $1\frac{1}{2}$ million, or double the figure for the late 1920s, and Harbron believes[1] that in 1937 nearly half of the foreign trade was carried in Russian vessels. The high average age and relatively small size of the vessels and the large proportion of outmoded coal burners, however, gave the fleet a poor level of efficiency, both technically and economically, and when the Second World War brought the need for a substantially increased inflow of supplies the U.S.S.R. became heavily dependent on the services of the Allied fleets.

The U.S.S.R. emerged from the Second World War with more tonnage than at the beginning—2·1 million tons in 1950, though this was only 20 per cent in excess of the 1913 tonnage. The remnants of the pre-war fleet were supplemented by about fifty obsolete U.S. freighters and about forty war-built U.S. Liberty ships (both groups retained although officially Lend-Lease material). In addition, some 180 ex-German vessels, which included relatively new freighters and large passenger ships, were handed over as reparations and a miscellaneous array of vessels was acquired after the defeat of Finland and the annexation of the Baltic States.

TRADE EXPANSION AND SHORTAGE OF SHIPPING, 1945–55

The need for ships in the late 1940s and early 1950s was greater than it had ever been, to bring in Polish coal, Romanian oil, Balkan ores, equipment and reparations from eastern Europe and the products of the 'Mixed Corporations', the

---

[1] Harbron, 1962, 140.

joint ventures established in all satellite countries except Poland and Czechoslovakia to produce a variety of manufactured goods with the Russian market as first priority. Indeed, so concerned was the U.S.S.R. about her vast requirements for reconstruction that, by 1952 (when Comecon was most effective in its exploitive phase), the U.S.S.R. took 58 per cent of Romanian exports, 57 per cent of Bulgarian, 48 per cent of East German, 35 per cent of Czechoslovakian, 32 per cent of Polish and 29 per cent of Hungarian, and there was a considerable exchange of goods with mainland China. The bulk of this substantial inflow had to be brought for at least part of the journey by sea; the east European railways were quite unable to accommodate such high-volume, long-distance flows.

At the same time, merchant shipbuilding in the U.S.S.R. was re-established surprisingly slowly. After the rebuilding of the damaged yards priority was clearly given to naval building, especially cruisers and submarines, and the first sea-going merchant ships do not appear to have been completed before 1953 when four 12 000-ton tankers were delivered.[1] The only known Soviet merchant ships completed 1945–55 appear to be twenty-four of this tanker class and two smaller tankers. New Russian-built dry cargo ships did not appear until 1956 and new fishing vessels until 1957.[2]

In such a situation in the decade 1945–55 *Sovtorgflot* was forced to look abroad for new tonnage, and she turned to Poland for the supply of fifty-two short-sea colliers and two cargo ships for the Mediterranean trade, to Finland for the supply of small coastwise tankers and general cargo vessels, to Denmark for refrigerated fish carriers and to Belgium for eleven fishing trawlers. Three polar research ships were also delivered during this period from the Netherlands.[3] These ship types are interesting, for they demonstrated some of the Soviet Union's very specialised maritime needs during the first post-war decade.

Clearly, however, these mixed and modest additions to the

---

[1] *Soviet merchant ships*, 1969, 8.
[2] Ibid., 1969, 9–11.
[3] Ibid., 1969, 6–11.

Soviet fleet were quite inadequate to handle the rapid expansion of her external trade which increased in value from 1·4 to 4·8 thousand millions of rubles between 1947 and 1952,[1] and it was in these circumstances that a threefold role was imposed upon the Polish economy, the major builder (for the time being) of new merchant ships for the U.S.S.R., as a significant contributor to the carriage of a growing volume of coal, ore and timber in the Baltic and North Sea trades (much of it directly or indirectly on Soviet account) and as an agency for the promotion of Soviet economic contacts beyond the confines of Europe.

By the acquisition of vessels from reparations and a variety of foreign sources, the Polish fleet stood at forty-six vessels of 160 000 tons gross at the end of the Three-Year Reconstruction Plan (1949)[2] and this appears to have been largely concerned with shipping coal and coke, which, in 1948, made up 12 million out of a total of 16 million tons of cargo shipments from Polish ports.[3] In such a situation it must have been exasperating for the Polish government to have been required to export (primarily to the U.S.S.R.) 136 of the 167 ships completed in the rebuilt Polish shipyards 1949–54.[4] By the end of the Polish Six-Year Plan (1950–5), however, the Polish tonnage increased by 70 per cent, and although a considerable proportion of the new ships were short-sea colliers and fishing vessels a growing fleet of dry-cargo ships was used to promote world-wide services, in particular to the Soviet Far East, North Korea and North Vietnam. Harbron provides evidence[5] of a substantial Polish share in the trade with mainland China in the 1950s, and vessels were supplied on charter to Sovtorgflot on a number of routes. As early as 1951 the P.A.K. Ocean Shipping Co. Ltd, a joint venture of Polish Ocean Lines and the Pakistan government, was established to link the two nations, with the liner *Batory* carrying passengers as well as goods.

East Germany's state merchant marine commenced operations in July 1952 and followed a not dissimilar growth pattern,

---

[1] Fairhall, 1971, 84.
[2] *Polish Maritime News*, No. 85, Sep. 1965, 2.
[3] Ibid., No. 81, May 1965, 9.
[4] Harbron, 1962, 85.   [5] Ibid., 1962, 85.

supplementing in the 1950s the other fleets in the Baltic, Balkan and Black Sea trades and contributing increasingly to North Sea trawling.

By 1955, the total deep-sea merchant shipping tonnage of the U.S.S.R., Eastern Europe and Mainland China represented only 3·3 per cent of the world's total volume of ocean-going tonnage and carried some 2 per cent of the world's seaborne cargoes, largely between the Socialist group of countries.

MARITIME EXPANSION IN ASSOCIATION WITH EAST EUROPEAN STATES, 1956–65

1956 constitutes an obvious watershed in this review for the revolts in Hungary and Poland in that year forced upon Comecon a more liberal economic policy for the satellite countries. Below-cost sales to the U.S.S.R. were removed, the 'Mixed Corporations' were phased out and state planning in eastern Europe was redesigned to give a greater measure of autonomous development with greater freedom for mutual economic aid and co-operation among the east European states. Of course, by this time, Soviet reconstruction was well advanced. Moreover, some of the rigidities of the international relations of the Stalin regime had been relaxed.

The immediate effects on the economies of Poland, East Germany and Czechoslovakia were striking and shipping played a most important role. Freed from the constraint to build primarily for the Soviet fleet, the Polish government invested 770 million zloty in shipyard expansion in the first Five-Year Plan (1956–60).[1] Output increased progressively from 120 000 to 174 000 tons a year and well over half of the new building now found its way into the Polish fleets. The traditional coal, timber and grain trades were supplemented by new cargo liner routes, to Japan in 1957, to West Africa in 1959, to India in 1960, to East Africa in 1962 and to South America in 1963; these routes supplemented the earlier ventures serving Pakistan, Canada and the Far East and the

---

[1] *Polish Maritime News*, No. 86, Oct. 1965, 16.

North Sea route to the United Kingdom. Indeed, at this time, a threefold stimulus caused Polish trading contacts with the non-Communist world to develop more quickly and more widely than Soviet contacts. In the first place, the new Comecon development policy generated a substantial transit trade through the three major Polish ports to and from Czechoslovakia, Hungary and parts of East Germany. This grew from 2·46 million tons handled in 1956 to 4·056 million tons in 1960.[1] Additionally, Polish ships commenced in 1958 a series of services for the carriage of Czech and Romanian goods from Black Sea ports to the Levant. In the second place, the Soviet government apparently continued to welcome the Polish merchant marine's ability to supplement Sovtorgflot's inadequate sea links with the Communist Far East, India and Latin America, to supply vessels on charter to the U.S.S.R. for bulk cargoes and to promote commercial links with the western world. In the third place, Polish maritime expertise and the integrity of her commercial transactions proved a valuable asset in the difficult circumstances of the east European development programme. Poland also contributed substantially in maritime fishery development, in particular with the design and production for the *Gryf* organisation, established in 1957, of mother ships which received at sea the catches of dependent trawlers, filleted, chilled, canned the catch and produced fish meal and cod liver oil. In 1959 the North Sea catch (77 400 tons) for the first time exceeded the Baltic catch (69 500 tons) and from 1961 'factory-trawlers' extended operations to the north-west Atlantic and the African continental shelf.[2]

A similar though less ambitious policy was followed in East Germany where from 1952 VEB Deutsche Seereederei came to share the Far Eastern, African and Mediterranean services with Polish Ocean Lines and a comparable deep-sea fishing fleet was rapidly expanded. In the sixties Romania and Bulgaria also acquired substantial fleets, initially contributing to the new trades with Cuba and North Vietnam.

Meanwhile in the U.S.S.R. from 1956–65 the merchant

---

[1] *Polish Maritime News*, No. 81, May 1965, 12.
[2] Ibid., No. 164, Apl. 1972, 18–20.

marine was built up from 2·5 to 8·2 million tons by the addition of an extraordinary array of about 830 vessels whose function was to some extent indicated by their type. Most numerous were general cargo vessels, relatively small in comparison with western fleets but with Russian-built types exceeding 8000 tons gross becoming more numerous after 1962. The bulk of these appear to have been shared between the short-sea trades (both domestic and international) in the Baltic and Black Seas, and the limited length of the average haul probably accounts for the small size of the ships used. Bakayev and Bayev state[1] that in 1960 only 15·6 per cent of the total domestic traffic was carried on the Pacific coast, only 4·2 per cent on the Arctic coast and only 5·3 per cent on the Baltic coast.

Second in importance in the decade 1956–65 came oil tankers, 186 in number, including forty-five exceeding 20 000 tons deadweight (large for the U.S.S.R., but of only intermediate size on a world standard). Most of them were based on the Black Sea and the landlocked Caspian, but the new Cuban market and the increased demand for shipments of crude oil, motor spirit and aviation fuels in the Soviet Far East, mainland China and North Vietnam placed particular strains on Soviet tanker capacity after the closure of the Suez Canal.

Several hundreds of fishing vessels formed the third group and these completed the planned establishment of a worldwide fishing industry, a complex and sophisticated operation quite separate from the other maritime affairs. Oceanographical research and reconnaissance vessels were built to explore and periodically review new oceanic fishing grounds. Mother factory ships, in effect floating canneries, were deployed in widely scattered areas and co-ordinated fleets of freezer trawlers provided the supply. The net catch increased from 2·6 to 6·1 million metric tons live weight from 1958 to 1968.[2] Most of it appears to have been delivered to domestic markets (the estimated annual U.S.S.R. demand for fish in 1964 was said to be 5·5–6·0 million tons)[3] but there is evidence that a

---

[1] Bakayev and Bayev (eds.) 1961, quoted by Alexandersson and Norstrom, 1963, 141.
[2] *Yearbook of Fishery Statistics*, F.A.O.
[3] Fairhall 1971, 159, citing T. Armstrong in *The Polar Record*, May 1956.

share of the catches was—and still is—collected by rendezvous at sea with Russian freighters and delivered directly to developing nations in Asia and Africa, some of it in the form of an economic aid programme.

The fourth group of ships in the 1956–65 decade consisted of passenger carriers. After the war the U.S.S.R. had acquired fifteen ex-German passenger ships, many of which appear to have been used in Pacific services, the largest (*Sovietskiy Soyuz*—ex *Hansa*) being on the regular run Vladivostok–Petropavlovsk. Others supplied cruises of varying degrees of sophistication from Soviet ports. Between 1959 and 1963, nine *Kirgizstan* type vessels (3220 tons only) were built in Leningrad, primarily for domestic Black Sea and Baltic routes. Over the same five years, nineteen larger (5260 tons gross), faster and more elegant ships, the *Kalinin* class, were built at Wismar in East Germany for the United Kingdom, Havana and other services. The most interesting innovation was the *Ivan Franko* class of some 20 000 tons gross, designed for international service. The prototype was delivered from Wismar in 1964 and spent her time on the U.S.S.R.–U.K.–Canada run or on cruising from Leningrad and Britain. Her consort, *Aleksandr Pushkin* (1965), followed a similar programme. These passenger ships also carried some freight and the *Kalinin* class built up a growing volume of cargo traffic on the North Sea route.

These developments emphasise the growth in volume and geographical dispersal of the U.S.S.R.'s external commitments at this time. Over the decade 1957–67 the value of all foreign trade increased from 7·5 to 16·4 million rubles. Exports by sea increased from 22 to 98 million tons, 1958–67, and this had become a more massive operation than the still-dispersed coastwise cabotage. Trade with other Socialist countries in eastern Europe and the Far East still created the greatest demand for shipping—some 70 per cent by value of the 1967 total, but trade with more recently aligned nations, particularly Cuba, India, China and Egypt, grew more rapidly and trade with the developed nations, though relatively small, increased at an even faster rate, the U.K., Japan, Italy, West Germany and France being the most significant partners. Thus, while the volume of foreign trade 1958–67 increased by 323 per cent

(26 to 110 million tons), the carrying capacity of Sovtorgflot (including fishery and research vessels) increased by only 200 per cent. Moreover, the efficiency of the fleet was greatly reduced by a serious unbalance between seaborne exports (85 per cent of all seaborne trade in 1958 and 90 per cent in 1967) over imports, thus necessitating many uneconomic home-bound voyages in ballast. It is not surprising, therefore, that of all foreign trade cargoes, 29·7 million tons (60 per cent of the total) were carried in foreign ships, most, but by no no means all, in east European ships.[1]

ACCELERATING GROWTH OF THE SOVIET MERCHANT FLEET SINCE 1965

It was this inadequacy of the Soviet fleet, coupled with the increased legitimate business of the east European fleets, the greater difficulty of chartering from the Soviet bloc currency area and the rising costs of chartering from other sources, that generated a further expansion of Soviet ship construction and purchasing in the late 1960s and early 1970s, and the scale of this was so substantial that considerable alarm was felt on both political and economic grounds in the western world. It is true, of course, that political objectives were encompassed within this most recent programme of planned expansion, but it is equally true that western observers in the late 1960s greatly underestimated the magnitude of the Soviet economic need for more shipping. Soviet trade and commerce were continuing to expand at a rate of 5·9 per cent annually (at constant prices) and growth rates in the associated east European economies were of a comparable order; Bulgaria 8·2 per cent per annum, Hungary 5 per cent, Poland 4·7 per cent, Eastern Germany 4·3 per cent and Romania 3·5 per cent.[2] Moreover, the increase in the volume of goods handled exceeded the increase in value. Over the nine years 1960–8, the average annual increase in goods loaded and unloaded at Soviet ports was 12·2 per cent while the comparable increase in the ports of Poland, Eastern Germany, Bulgaria and

---

[1] Bakayev, 1969.
[2] *Yearbook of international trade statistics*, U.N.

Romania combined was 9·3 per cent,[1] the latter growth being progressively inflated by a growing transit trade to and from the land-locked states of Czechoslovakia and Hungary. The seaborne transit traffic that passed through Polish ports (most of it moving between central Europe and the U.S.S.R.), for instance, grew from 4·05 million metric tons in 1960 to 6·75 million in 1968 and 8·07 million in 1971.[2] A comparable growth of traffic originating in central European ports on the Danube provided a continuous stimulus to the Black Sea trade routes. Moreover, although the eastern European fleets involved in these trades were being expanded, Sovtorgflot aspired to maintain a leading role.

An equally potent factor in the demand for more Soviet shipping tonnage was the increasing length of the average voyage which had more than doubled from 935 miles in 1960 to 2161 miles in 1968.[3] Fairhall states[4] that sixty to sixty-five Soviet vessels were required for the regular U.S.S.R.–Cuba run in 1969. The closure of the Suez Canal had significantly added to the length of the haul between the U.S.S.R. and India, North Vietnam, China, North Korea and the Soviet Far East. Purchases of wheat and wool took Russian cargo ships to Australia and Canada, and Sovtorgflot had a significant involvement in transporting the massive Chinese purchases of grain from these countries. The success of these transactions, coupled with the growing Soviet demand for foreign raw materials, brought by her steadily increasing population, rising living standards and the consequential need for foreign currency earnings, motivated a growing array of trading negotiations beyond the bounds of the aligned nations. These have matured in a variety of forms. The U.S.S.R.–Japanese trade agreement of September 1971 concerned the mutual exchange of a variety of goods; the 1972 commodity agreement with Brazil concerned the purchase and transport of a reported 290000 tons of sugar by the U.S.S.R.—over 300000 tons of sugar were also stated to have been purchased by China and despite the strains on Soviet–

---

[1] *Yearbook of international trade statistics*, U.N.
[2] *Polish Maritime News*, No. 165, May 1972, 9.
[3] Fairhall, 1971, 78.
[4] Ibid., 73.

Chinese relations Soviet vessels have probably been involved in the transfer.¹ In July 1972 a different type of contract was signed with the Chilean government which awarded to the Baltic Steamship Company the charter to ship 190 000 tons of copper exports (by no means all of it to the U.S.S.R. market).² The most massive of the commodity agreements, however, involved the purchase of some 19 million tons of wheat, coarse grains and soya beans from the United States, valued at $750 millions in 1972. The details of this complex deal took nine months of negotiation and embraced the share of the traffic to be carried in American, Soviet and other ships, the basis for agreement on freight rates and the conditions on which Soviet ships were to be permitted entry into specified American ports and American ships could have access to Soviet ports.³ Even the terms of settlement of the U.S.S.R.'s residual Lend-Lease debts required resolution before the final shape of the Soviet-American maritime agreement emerged in November 1972. Political considerations and consequential benefits apart, the purely financial stakes in this deal were very great. The first shipments in the late summer of 1972 were booked at a freight rate of $5·80 a ton. The market rate rose rapidly to $7·50 a ton while negotiations continued through the autumn.⁴ The final cost of shipping the entire purchase could well exceed $200 million and the Soviet determination that at least one-third of this should be paid for in rubles rather than dollars accounts for the tenacity of their bargaining. Equally interesting was the American insistence that a third of the grain should be shipped in U.S. vessels, even though this would involve a subsidy of $2 or more for every ton shipped ($12 million in all) from the U.S. Federal Maritime Administration to meet the difference between the official charter rate for the American vessels and the rate paid by the Russians.

The costings of this particular transaction effectively illustrate the financial value of maximising the use of Soviet shipping to carry the growing volume of freight to and from

---

[1] *Fairplay*, No. 4649, 28 Sep., 1972, 15.
[2] *Seatrade*, Vol. 2, No. 12, Dec. 1972, 75.
[3] *Fairplay*, No. 4658, 30 Nov. 1972, 15.
[4] Ibid.

Soviet ports. They illustrate equally cogently the rewards of building up shipping connections for the 'cross-trades' in which the cargoes carried are the merchandise of other nations and the freight rates received represent substantial earnings in foreign currencies. The advantages of such arrangements were emphasised by the chagrin of Chile's major shipping line when the ore contract previously quoted was signed and by the strenuous opposition from the U.K./Continent—Australia conference lines when the U.S.S.R. insisted in 1968 not only that Soviet purchases of Australian wool should be carried in Soviet vessels but that Sovtorgflot would also compete at below-conference tariffs in the general trade between Europe and Australia. After long and aggressive negotiations the Baltic Steamship Company in 1969 joined the Europe to Australia and Europe to New Zealand shipping conferences, operating in co-ordinated competition with western lines and charging the same rates.

It is evident, therefore, that the most recent programme of maritime expansion was aimed at a widespread competitive involvement in international trade routes as well as the provision of a more complete service for Soviet and Comecon overseas trade and this new policy brought with it two implications of great significance.

In the first place, the unbalance between ships moving fully laden outbound and only partially laden inbound proved to be as unsatisfactory with the new services bringing Brazilian sugar, Chilean minerals, Australian wool and American grain to Europe and the U.S.S.R. as it had been with the established services to the Far East and India. The most effective means of minimising such disparities was to adopt the long-established west European and Japanese practice of 'triangular' route patterns. Within the socialist realm the Polish and Yugoslav fleets had provided the model answer in the mid-1960s—the despatch of ships fully laden to the Far East, the carriage of mixed cargoes thence to Australia where the ships were conveniently placed for the homebound voyage carrying grain and wool. The U.S.S.R. experimented with this practice after the first substantial purchases of Australian wheat in the late 1960s only to find, hardly unexpectedly, that Australian waterside workers strongly objected to loading Soviet ships newly arrived from unloading military stores at Haiphong, and more devious route-plans had to be devised. The deviation from direct liner cargo voyages between the U.S.S.R. and Cuba to deliver Soviet exports in Latin America before returning with Cuban sugar was more successful, and joint cargo liner services to India, Ceylon and the Persian Gulf brought convenient ship placements for other cross-trades.

Such strategies in themselves added to the need for new tonnage. Even more important, they created the need for new and more specialised ships—and above all for larger ships— than Sovtorgflot and its associates had been accustomed to using. In the 1960s rising costs and cut-throat competition in the western world had forced a technical revolution in merchant ship design and operation. Since the mid-1960s the largest tankers have increased in size from 100 000 tons deadweight capacity to 275 000 tons, and half-million tonners will be available in the near future. The diseconomies of voyages in ballast have forced the design of combination carriers, the so-called 'OBOs' which can carry dry cargoes in bulk inbound and fluid cargoes outbound. Rising crew costs have brought the demand for automated operation and 'unmanned' engine rooms. Rising waterfront labour costs gave birth to the containerisation of cargoes and the introduction of 'roll-on' vessels whose freight could be driven on board, and of large barge-carriers each carrying eighty or more large self-propelled lighters which can be offloaded by heavy gantries and proceed independently coastwise or up-river. Such measures have greatly increased the efficiency of operation but also have greatly added to the capital investment required for new tonnage and port modernisation.

The U.S.S.R. had notably lagged behind in each of these areas of technical development. Labour-intensive manning was not such a problem as cost rises were less in the U.S.S.R. than in the west and the economics of subsidy were probably so camouflaged that the diseconomies of large crews and waterfront teams were not so apparent. Moreover, the navigational limitations of the Danish straits and the Bosporus and therefore also of the accommodation for large ships in Baltic and Black Sea ports placed constraints on the maximum size of vessels that could be used.

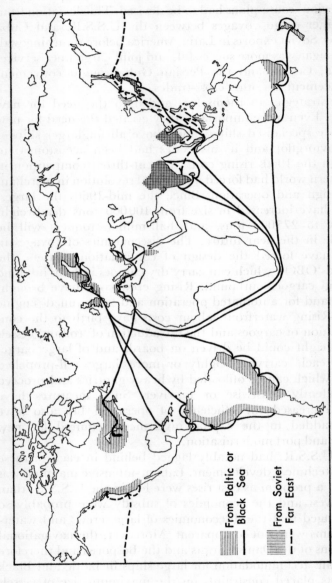

Fig. 10. The major regular cargo-liner services operated by the U.S.S.R. and the approximate areas served, 1972. Services administered by the Far East Maritime Shipping Company are shown in pecked lines.

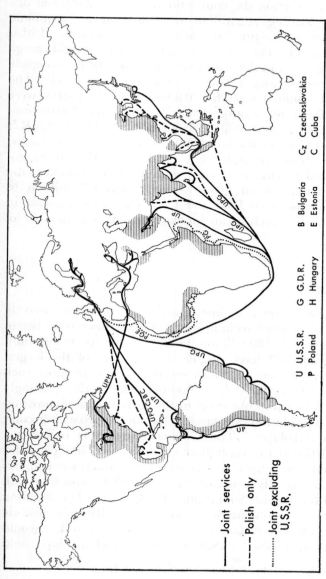

Fig. 11. The major Comecon joint cargo-liner routes and other routes served by Polish Ocean Lines, 1972. The approximate extent of the areas served is shown by the horizontal ruling. Coastwise services in the Baltic, Black Sea and the North Sea are omitted.

Continuous lines indicate Comecon joint services with Soviet participation. Dotted lines indicate Comecon joint services without Soviet participation. Pecked lines indicate other services operated by Polish Ocean Lines. *Note*: Soviet services from Europe to Australia and New Zealand and from Soviet Pacific ports are excluded (see Fig. 10 for details).

In such a situation the conventional Soviet 12 000-ton dry cargo ship loading Australian wool was at an economic disadvantage in comparison with the 30 000-ton British container vessel. The medium-sized Soviet general cargo ship loading grain in American Gulf ports might lie alongside a Dutch 100 000 tonner competing in the same trade. The operational economics of the 20 000-ton Soviet tanker were notably inferior on transoceanic voyages to the 250 000-ton Greek, American and Scandinavian tankers. The U.S.S.R.'s most urgent need in the late 1960s, therefore, was for larger, faster and more versatile ships with sophisticated cargo-handling and machinery-control equipment. And indeed, the remarkable increase in gross tonnage from 8·2 million to 16·04 million between 1965 and 1972 reflects the increase in size per unit as much as an increase in the total number of units.

OUTLOOK

The major Soviet maritime innovations (or more correctly adoptions of western techniques), however, have only begun very recently. A 180 000-ton tanker, currently being built at Leningrad, will have double the capacity of the largest oil carrier at present in Soviet service, and three more such vessels are on order.[1] A 63 000-ton ore carrier (wrongly identified by western observers in the early stages of construction as a possible aircraft carrier)[2] is now approaching completion at Nikolayev.[3] The Gdynia shipyard is building a series of 105 000 ton oil-bulk-ore carriers and has plans for 200 000 tonners.[4] One can only conjecture about the employment of such vessels assuming that the 100 000-ton limit still applies to the Danish Sound and the Bosporus. The situation emphasises the contemporary version of Russia's historic search for wholly satisfactory maritime outlets, and it would indeed be ironic if the U.S.S.R.'s prestige merchant ships could

---

[1] World ships on order, *Fairplay*, No. 34, 22 Sep. 1973.
[2] *Time*, 31 Dec. 1972, 10.
[3] *Seatrade*, Vol. 2, No. 4, Apl. 1972, 49.
[4] *Fairplay*, No. 4642, 10 Aug. 1972, 27.

not reach European Soviet ports in a fully laden state. Shipment of bulk cargoes from Latin America and Australia to the U.S.S.R. may be practicable if part of the cargoes are off-loaded in E.E.C. or Mediterranean ports en route. Employment in cross-trades beyond the confines of the Baltic and Black Seas and in competition with established international consortia is likely to be much more profitable when the large Soviet bulk carriers are fully operational. The U.S.S.R.'s April 1972 trade agreement with Iraq may well facilitate the shipment of oil in Soviet tankers to Japan, and the decision to ship urea fertiliser in large quantities to China from the huge petrochemical plant at El Tablazo in Venezuela may provide yet another novel opportunity for Soviet shipping.[1]

In the container-shipping field the Soviet Far East has proved a profitable area for experimentation. Motivated by current Japanese successes in 'unitised' transport methods, the Soviet Far East Maritime Shipping Company (FESCO) introduced container carriers on the Nakhodka–Japan route in 1970. Heavy-duty flat trucks were then introduced to the Trans-Siberian Railway to provide through container transport from Moscow to Tokyo, and in 1972 the London–Leningrad ships offered a container service preparatory to establishing the much publicised through service London to Tokyo via Moscow. In the six months to June 1972 it was reported[2] that nearly 5000 containers were moved between Europe and Japan via the Trans-Siberian Railway and that 1750 of them originated in western Europe. The annual flow is programmed at 15–20 000 containers for the year 1973, and it is interesting that 1000 of the new containers required by the Anglo–Soviet Shipping Company were ordered from a Cheshire manufacturer.[3] Transhipment at the Pacific terminal has been facilitated by the construction of costly new port facilities at Nakhodka and a new harbour at Vanino to cater for the ferry service to Sakhalin, and an array of feeder services to Korea, South Japan and Hong Kong is being 'containerised'. To

---

[1] *Seatrade*, Vol. 3, No. 1, Jan. 1973, 23.
[2] *Fairplay*, No. 4659, 7 Dec. 1972, 13.
[3] Ibid. 4663, 4 Jan. 1973, 30.

reduce the directional unbalance of traffic on these services, rates at a 20 per cent discount compared with the conference lines operating on conventional all-sea routes have been offered to shippers in south-east Asia in respect of goods shipped to Europe via Nakhodka and Moscow.[1] A comparable proliferation of container services between Black Sea and Danubian ports has been designed to feed into the trans-continental rail system from the south-west.

The continuous modernisation of the Soviet rail system to provide integrated links with their shipping services has retarded the growth of international road transport services for the carriage of goods. For this reason the road traffic 'roll-on' shipping which has been so extensively adopted for long-distance services by Scandinavian shipping companies has not hitherto found favour in the U.S.S.R. In 1972, however, orders were placed for two combination 'roll-on' container vessels from the Finnish Valmet yard, virtually identical to ships from the same yards for Scandinavian owners which were designed for the Australia–North America service. *Seatrade* journal reported[2] thirteen other 'roll-on' ships on order in west European shipyards for the U.S.S.R. The placement of these vessels, and of the Polish refrigerated B 443 series designed for the carriage of fruit, butter and meat,[3] will be interesting to know.

In the late 1960s, when western shipping companies were withdrawing passenger vessels from liner services in the face of airline competition and finding that the steeply rising costs were eroding profit margins on the operation of cruise-ships the U.S.S.R. was, for the first time, introducing the new 20 000-ton 'poet' class of passenger vessel into service. Heavily subsidised, intended for the promotion of international public relations and as earners of foreign currencies, they filled the vacuum created by the contraction of western passenger shipping services. The *Ivan Franko* has at times been the only passenger vessel on the U.K.–Canada run. The *Shota Rustaveli* has had charter contracts to carry U.K. and west European

---

[1] *Seatrade*, vol. 3, No. 1, Jan. 1973, 41.
[2] Ibid., vol. 2, No. 7, July 1972, 57.
[3] *Polish Maritime News*, No. 155/156, July/Aug. 1971.

migrants to Australia. Both vessels have operated cruise services out of British ports. The Baltic Steamship Company's British cruising programme has reached a peak with the completion of the fifth ship, the *Mikhail Lermontov*, in April 1972. All five vessels are offering a total of twenty-three cruises from Tilbury or Southampton in addition to nine fly-cruise tours based on Gibraltar in 1973.[1] These vessels offer an element of novelty as well as an economical cruise in an elegant vessel and the operation has now become much more than a flirtation with the British cruise market.

Looking at the contemporary Soviet shipping scene from a world perspective, one can have no doubt about the scale and the aggressiveness of the worldwide commercial operations of Soviet shipping that are being extended into the future with continued vigour. *Fairplay*'s August 1972 supplement[2] recorded as being on order for the U.S.S.R. at 31st July 1972

334 dry cargo ships of 1·872 million tons deadweight
35 container ships of 0·486 million tons deadweight
103 tankers of 1·079 million tons
17 bulk carriers of 0·569 million tons.

Substantial though they appear, these orders represented only 2–3 per cent of total world orders and their fulfilment over the next three to four years would not improve the Soviet's ranking behind the fleets of Japan, Liberia, the U.K., Norway, the U.S.A. or the E.E.C. countries. Even if one regards the combined fleets operated by Comecon as a single unit, the aggregate tonnage is likely to fall short of that of their major 'western' competitors, and the U.S.S.R.'s dominant position in the orders for dry cargo ships and intermediate sized tankers owes much to the fact that these classes are not the prime choice of 'western' operators who prefer bulk, container or large tanker vessels. Although, therefore, the U.S.S.R.'s growth to sixth place among the ship-operating nations creates another parameter in the competitive universe of world

---

[1] *Seatrade*, vol. 3, No. 1. Jan. 1973, 17.
[2] 'World ships on order', *Fairplay*, No. 32, 24 Aug. 1972.

shipping, it is an inevitable response to the continued growth of the centrally planned economies, and it is significant that despite this growth the U.S.S.R. shows no signs of losing its dependence on chartered foreign tonnage. A leader in *Fairplay* (10 August 1972) entitled 'Cheerless apart from the Russians', lamenting the general dearth of charter contracts for shipping in the world as a whole, quoted the booking of 400 000 tons of chartered tonnage to bring $3\frac{1}{2}$ million tons of North American grain to the Soviet Union in the late summer of 1972. The Russians, it seems, like Alice, have to run as fast as they can to stay in the same place.

BIBLIOGRAPHY

Alexandersson, G. and Norstrom, G., *World shipping*, New York, 1963
Athay, R. E., *The economics of Soviet merchant-shipping policy*, Chapel Hill, 1971
Bakayev, V. G. and Bayev, S. M. (eds.), *Transport SSSR*, Moscow, 1961
Bakayev, V. G., *Soviet ships on world sea routes*, Moscow, 1969
Food and Agriculture Organisation of the United Nations, *Yearbook of fishery statistics*, annually
Fairhall, D., *Russia looks to the sea*, London, 1971
Fairplay International Shipping Journal, London, weekly
Harbron, J. D., *Communist ships and shipping*, London, 1962
Nadtochiy, G. L., *Geografiya morskikh putey*, Moscow, 1972
Polish Maritime News, Gdynia, monthly
Rudoi, Y. and Lazarenko, T., *Transport and communication 1959–65*, Moscow, n.d.
Seatrade, Colchester, monthly
Stamp, L. D., *Intermediate commercial geography*, London, 1933
Time, Chicago, weekly
United Nations *Yearbook of international trade statistics*, New York, annually
Schwarz, H., *Russia's soviet economy*, Englewood Cliffs, 1954
Soviet merchant ships, anon., Havant, 1969
Zaleskiy, E. *Geografiya morskogo transporta*, Moscow, 1971

5

The Northern Sea Route

The northern sea route, or the north-east passage as it used to be called, is the sea route round the north of Asia. It may be taken as beginning at the straits which lead from the Barents Sea to the Kara Sea. The Barents Sea itself and the White Sea which leads off it are not properly part of the route, even though the main western termini, Murmansk and Arkhangelsk, face those seas. The route may be said to end on the Pacific side at Bering Strait, but the eastern terminus is Vladivostok, far to the south-west (Fig. 12). A study of the present state of development of this route, and of the part it plays in the economy of the Soviet Union, is clearly not central to any study of the Soviet transport network as a whole. It does show, however, in a way in which other studies might not, how the Soviet system reacts to a particular challenge in a field in which it cannot be guided by the experience of others, but must itself be the pioneer.

There have been histories of the northern sea route. The most compendious and the most authoritative (although displaying strong nationalist feeling) is the four-volume *Istoriya otkrytiya i osvoyeniya severnogo morskogo puti* (Moscow 1956–69), by M. I. Belov and (for volume 2) D. M. Pinkhenson. This valuable work takes the story to 1945. Shorter works in English are C. Krypton's *The northern sea route. Its place in Russian economic history before 1917* (New York, 1953), and the same author's companion volume *The northern sea route and the economy of the Soviet north* (New York and London, 1956); and the present writer's *The northern sea route, Soviet exploitation of*

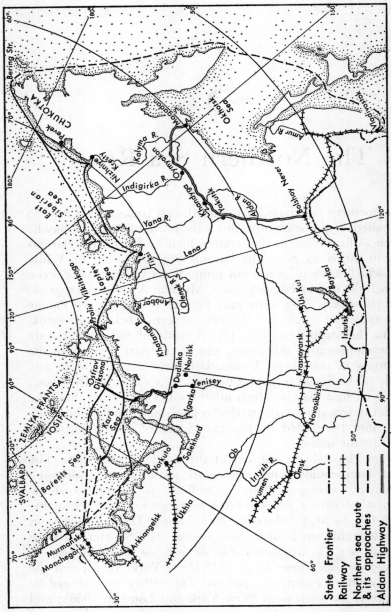

Fig. 12. The northern sea route and principal transport routes in Siberia and the Soviet Far East.

THE NORTHERN SEA ROUTE

the north-east passage (Cambridge, 1952). These provide much of the background, and it is unnecessary here to summarise that material again. Our object now is to examine the present working of the route. The source material for this is unfortunately scanty. No even moderately full account of the operations of even a single year since 1945 has appeared in the Soviet Union. All that has appeared has been occasional news items in the national or specialist press. What follows, therefore, is an attempt to construct a coherent picture out of these relatively haphazard pieces of news reporting. The main sources are *Pravda, Vodnyy Transport, Morskoy Flot* and Moscow Radio as monitored by the B.B.C. and published in *Summary of World Broadcasts*.

ICE

The major obstacle to shipping in these waters is the presence of floating ice. There are two main kinds: first, the frozen surface of the water body, called sea ice when the water body is the sea; and second, ice which was formed on land, particularly in glaciers, and has reached the sea where it has broken away from its parent mass and floated off as an iceberg. This second kind, which is the more spectacular, accounts for much less than five per cent of floating ice, and is found so rarely on the northern sea route that it can be ignored. Sea ice is the only problem.

The extent of sea ice is highly variable. For most of the year the whole of the northern sea route is covered by it. For a period in the summer, however, the whole of it is ice-free—though not necessarily all sections of it for the same period. This seasonal variation is paralleled by a smaller but still considerable variation between the maxima and minima of different years. Dates of break-up and freeze-up can vary very much from year to year. When the ice has broken up and melted in one place this does not mean that that place will remain ice-free until freeze-up; most of the route is fully open to the north, and a northerly wind may at any time bring down heavy ice from the central Arctic Ocean, which does not clear in the summer. Thus ice *may* be encountered on almost any part of the route at almost any time of the year, and *will* be encountered on all parts of the route at most times of the year.

The ice will be of varying thickness and strength. In the course of its first winter of growth it may attain a thickness of up to two metres, if undisturbed. If the floes have been broken by pressure into ridges and hummocks, a process which will have occurred in many areas, the consolidated fragments may reach thicknesses of ten metres or more. This first-year ice, as it is called, is the predominant ice type encountered, but the ice which has been blown down from the north is likely to be older, having survived into a second or later winter. This ice, known as old ice, is thicker; a level floe may reach three metres, but most floes are rather intensively ridged and hummocked and this may produce thicknesses of twenty metres or more. Such ice is also disproportionately stronger, because most of the brine has by now been leached out. By and large, it is the inclusions of old ice in fields of first-year ice which cause difficulty for ships.

It must be remembered that most of this sea ice is in continuous movement: cracks between floes open up to form leads, leads freeze over with new ice, floes collide again to form ridges. There is also a general pattern of drift. North of the northern sea route the ice drifts mostly in a westerly direction, towards Spitsbergen (Svalbard) and the Greenland Sea, where it is ultimately borne out into the North Atlantic to melt. Some of the ice is not in movement, however. This is the so-called fast ice which remains frozen to the shore or held in position by off-lying islands. In parts of the Laptev and East Siberian Seas this fringe of fast ice attains a greater width than anywhere else in the Arctic—up to 500 km from the mainland. All this is first-year ice, clearing away completely in the summer.

FREIGHTING OPERATIONS

There is not enough information on which to reconstruct the events of, say, the 1972 season, or of any other single recent season. Consideration of news releases over the last three or four seasons, however, does permit one to make a composite guess as to what happens each year. The number of freighters employed is of the order of two to four hundred, and the total freight lifted of the order of one to two million metric tons. The traffic breaks down into several component parts.

The largest single freighting movement is that between the western termini and the river Yenisey (Fig. 12). This river has depths permitting ocean-going ships to reach as far upstream as Igarka, a timber-milling centre 673 km from the mouth. Downstream from Igarka is the river port of Dudinka, from which a railway runs the 96 km to Norilsk, an important nickel and copper mining centre with a population in 1970 of 136 000. Igarka generates freight in the form of timber, and this traffic is currently running at about 220 000 standards (650 000 metric tons) a year, carried in over 100 ships. There is no appreciable inward freight to Igarka across the Kara Sea, since what is required can come more easily down the river. Dudinka, however, does take general cargo for Norilsk, and also handles ore leaving Norilsk. This last has become more important recently as the nickel smelter at Norilsk is overloaded and there is surplus capacity at Monchegorsk, south of Murmansk. To what extent it has been possible to balance up this traffic by putting freight bound for Norilsk into the ships coming for timber to Igarka is not clear. Exports from the Yenisey certainly exceed imports.

Analogous to the Yenisey traffic, but on a much smaller scale, is the traffic from the eastern terminus to the port of Pevek in Chukotka, and to Nizhniye Kresty at the mouth of the Kolyma river. These two ports serve goldmining areas. The freight is therefore fuel and general cargo taken into the ports, with very little for the back-haul. Exceptionally, this traffic has employed as many as 130 ships (in 1961), when it exceeded the Yenisey traffic in volume, but normally it uses probably less than half this number. There is no indication of the tonnage carried, but it might be of the order of 300 000 metric tons a year.

The port of Tiksi, at the mouth of the Lena river, being roughly mid-way along the northern sea route, is commonly reached from both ends. It is the transhipment point for the Lena, which will not admit ocean-going vessels. It is also the centre from which coastwise traffic operates to the Khatanga, Olenek, Anabar, Yana and Indigirka rivers. In the spring the great volume of relatively warm river water melts the ice offshore in the region of the Lena delta before the rest of the sea ice is melted, so operations can start earlier than the arrival of the first ships from outside. To help in this a small icebreaker often overwinters at Tiksi.

A fourth freighting operation is the annual re-supply of a number of small stations—scientific, commercial and no doubt military—scattered along the coast and on the offshore islands. This probably does not involve lifting a large tonnage but there cannot be less than a hundred such stations along the route.

One particular convoy has been repeated each season since 1948. It is the passage of river craft—barges, tugs, passenger boats, dredgers and floating cranes—from the White Sea to the Siberian rivers on which they are to work. These ships have been built in European U.S.S.R. or eastern Europe and the northern sea route is the only way to transport them. The convoy, of as many as seventy-six ships, moves eastwards along the coast and ships break off as their river of destination is reached. The voyage cannot be easy for the ships are not built to withstand either heavy ice or heavy seas. No disasters have been reported, but disasters are not news in the Soviet Union. A more reliable indication that these expeditions have been successful is the remarkable fact that the same man, F. V. Nayanov, was in charge of the convoy every year up to 1967 at least, and probably longer (he was active in 1972, the twenty-fifth consecutive season).

Finally, there is the small but growing tourist traffic to these waters. A Soviet cruise ship, *Vatslav Vorovskiy*, has made a voyage each summer in recent years from Murmansk to Zemlya Frantsa Iosifa, and thence to Ostrov Diksona, the Yenisey estuary, and back to Murmansk. A less ambitious tourist route is provided by river steamers on the Yenisey which ply from Krasnoyarsk down to Dikson and back. There have even been a couple of sporting voyages along the northern sea route in sailing boats.

A notable feature of this traffic has been the very small part played by through voyages from one end of the route to

THE NORTHERN SEA ROUTE

the other. When the Soviet government decided, in the early 1930s, to put a big effort into making the route usable there is no doubt that strategic considerations played a part in the decision. Japan was seen as a likely opponent, and memories were still fresh of the Russo-Japanese war in which the Russian fleet was defeated at Tsushima largely because the Japanese knew it was coming and were fully prepared. The northern sea route offered a much more secure way of transferring ships betwen the Atlantic and the Pacific. It is probable that use was made of this during the Second World War, although it could never have played a critical part because the Soviet Union was at war with Japan for only six days in 1945. Possibly there has been further naval use since that time, but economic use has been at an insignificant level. Ships which have made the traverse have generally done so by chance; ice has closed in on their return route, obliging them to withdraw the other way. The northern sea route as complement to the overloaded Trans-Siberian Railway has not worked out. Clearly the short season and the possibility of delay or failure even during that season are powerful deterrents.

SHIPS

The prime requirement for making these voyages is a fleet of vessels able to operate in ice. Two main types are needed; icebreakers, which carry no cargo but break a track for ships which do; and ice-strengthened freighters, which can deal with some ice unaided but must be escorted through heavy concentrations.

The Soviet Union employs on the northern sea route about twenty icebreakers, of which a dozen are of 10 000 h.p. or more, and the rest are 'port' type icebreakers of about 5000 h.p. The bigger ships are mostly Finnish-built, by the firm OY Wärtsilä AB of Helsinki. These are the five *Moskva*-class of 22 000 h.p., built 1960–9, and the three *Kapitan*-class of 10 500 h.p., built 1954–6. The rest are Soviet-built: all the smaller icebreakers, two surviving *Stalin*-class of 10 000 h.p., built 1938–41, and finally the flagship of the fleet, the nuclear-powered *Lenin* of 44 000 h.p., built 1959. *Lenin* was the first example of nuclear propulsion in any surface ship and she has performed some remarkable feats. In particular, she made a voyage in October–November 1961 to the northern parts of the East Siberian and Laptev Seas, but she was out of action from 1967 to 1970, probably with reactor trouble.

There were plans, first announced in 1964, to build more nuclear-powered icebreakers in the Soviet Union. With the return of *Lenin* to active service in 1971 more has been heard of these plans. Two—some reports say three—more ships are to be built. The first, called *Arktika*, is to be completed in 1975. These ships are to be nearly twice as large and powerful as *Lenin* (25 000 metric tons displacement to *Lenin*'s 16 000, 75 000 h.p. to *Lenin*'s 44 000). Meanwhile, orders for three ships in a new class of 36 000 h.p. have been placed in Finland. The first two, called *Yermak* and *Admiral Makarov*, are due for completion in 1974 and 1975, and the third in 1976. Thus, by the late 1970s there will be very powerful reinforcement of the Soviet icebreaker fleet—a fact which reflects not only confidence in future operations but an evident desire to expand them.

The method of breaking ice employed by these ships remains the traditional one of mounting and crushing with the ship's heavily reinforced stem. Other methods have often been discussed, but up to now none has reached the stage of operational use. It is amusing to recall that the idea of the 'Alexbow', a means of breaking ice from underneath tested experimentally in Canada in 1967–8, was investigated and evidently discarded by the Russians in 1940–1. The scattering of dark powder on the ice surface, causing an increase in radiant heat, has been used as a way of weakening the ice before it is broken. This method has been applied in the estuary of the Yenisey, apparently with some success.

Most of the freighters employed on the northern sea route are strengthened for ice. Of the five classes in the Soviet ice classification for freighters (icebreakers are in a class by themselves), the top two permit operation in the Arctic. The Soviet Union has over 800 ships in these two classes. They are of rather small cargo capacity, most being within the range 3 000 to 9 500 tons deadweight, presumably due to restrictions imposed by shallow water. Some container ships are now in use. Two of 13 000 tons deadweight (the largest ships known to have used the route) have been employed on the Arkhangelsk–Dudinka run since 1967, each ship making two round trips a season. In the opinion of Soviet specialists more ships of this type should be built, but somewhat smaller. The latest development has been the introduction of what is described as a new type of freighter able to operate in both sea and rivers. One of this class, *Morskoy-21*, reached the Yenisey in 1971. In what way it differs from those freighters which habitually make the voyage up the Yenisey to Dudinka and Igarka is not yet apparent, but the designers were awarded a state prize in 1972.

The crewing of these ships is a point of considerable importance. Navigation in ice-filled waters is a skill which most seamen never acquire, and indeed their training is to avoid ice at all costs. A skilled crew under expert and determined leadership might take a vessel through, when the same vessel in different hands might be obliged to turn back or even to winter at sea. It is fair to say that at least fifty ships have been employed every season on the northern sea route since 1935, so not only have many persons acquired great experience but a career structure in Arctic seafaring has been in existence for more than a generation. This is probably one of the most important factors in the successes which have been achieved.

LENGTH OF SEASON

The concept of the length of the operating season on the northern sea route is necessarily complex. The length varies from place to place on the route, and from year to year. It also varies with the type of ship employed and the determination of the captain. Assuming an ice-strengthened ship with an experienced captain, the availability of icebreaker support on call and regular access to synoptic reports and forecasts of the ice situation, one can say that the Ob and Yenisey are usually accessible from the west between early July and early November; the Kolyma and Pevek from the east between mid-July and late October; the Lena from either end between early August and mid-October. Thus, ports only a short distance along the route may have a season of up to four and a half months, while the through route has perhaps two and a half.

An interesting development has taken place in recent seasons. In 1970 an experimental late voyage was made from the west to the Yenisey and back by a freighter with icebreaker escort. This was successfully completed between 15 November and 3 December. A more ambitious operation was carried out in 1971, when six freighters, escorted by three icebreakers, followed the same route still later in the season, the last ship leaving the Yenisey on 26 December. In 1972 nine freighters, escorted by five icebreakers, operated on the same route between early December and late January (1973). If such voyages become routine the season for this part of the route will be effectively lengthened to six months.

ADMINISTRATION

Operations and services on the northern sea route come under the control of the Ministry of the Merchant Fleet. In the 1930s there was greater independence, control being vested in *Glavsevmorput'* (Chief Administration of the Northern Sea Route), which was at its creation in 1932 independent of any ministry, but subordinated directly to the Council of Ministers (or Council of People's Commissars, as it then was). Glavsevmorput had very extensive responsibilities, not only for the sea route, but also on land. These last were greatly curtailed in 1939–41. In 1953 Glavsevmorput was placed under the wing first of a combined Ministry of the Merchant Fleet and the River Fleet, and then of the Ministry of the Merchant Fleet when that became a separate entity again in 1954. Glavsevmorput thus became simply the Arctic shipping division of the Ministry, and in 1963 the name itself was dropped, presumably indicating that complete absorption had taken place.

On 16 September 1971, however, legislation was passed by the Council of Ministers of the U.S.S.R. setting up a new Administration of the Northern Sea Route attached to the Ministry of the Merchant Fleet. There has been no mention of its activities in the press, and it is not clear just what its role will be. Although its name evokes that of Glavsevmorput, it does not appear to be a revival of the earlier organisation in any closely similar form. Glavsevmorput was a Chief Administration, which is higher in the hierarchy of government offices than an Administration. The Soviet national press (*Pravda*, 19 December 1972) mentioned the fortieth anniversary of the creation of Glavsevmorput, but made no reference to the Administration of the Northern Sea Route. The new organisation has as its main objectives supervising the rational use of the northern sea route, organising navigation on it and preventing pollution—a narrower mandate than Glavsevmorput had. It is likely that the Canadian anti-pollution legislation of 1970 played a part in the formulation of the new statute. The full text is given in translation in *Polar Record*, vol. 16, No. 102, 1972, 419–21.

INTERNATIONAL USE

By far the major proportion of ships using the northern sea route have been Soviet. Non-Soviet ships have sailed these waters only in particular circumstances. The commonest is the charter of Norwegian, British and other freighters to transport timber from Igarka. These ships are under quite close Soviet control for the Arctic section of their voyage. In the early years most of the ships engaged in this trade were foreign, but from the 1950s the proportion has dropped as the Soviet merchant fleet has grown. This is the only freighting operation which employs any foreign ships. A quite different type of foreign use, by ships not under Soviet control in any way, has been the series of voyages by United States Coast Guard icebreakers. In each season during the years 1962–7 one or sometimes two of these ships made oceanographical cruises to one or more of the constituent seas of the route. These ships were under constant surveillance from both air and sea by the Soviet armed forces and it was made quite clear that their presence was unwelcome, but no attempt was made to expel them so long as they remained outside territorial waters. However, when two of them tried in 1967 to enter Proliv

Vilkitskogo, the strait joining the Kara Sea and the Laptev Sea, the Soviet government objected because the strait is twenty-two nautical miles wide at its narrowest point. The ships turned back into the Kara Sea. The United States government protested at denial of the right of innocent passage, to which the Soviet answer was that warships (the vessels carried machine guns) should give thirty days' notice and this had not been done. The matter could have been argued further, but clearly this was a bad place and time for the United States to pick a quarrel. The important point here is that, despite strong earlier statements by Soviet jurists, the Soviet government did not seek to claim any jurisdiction over the high seas in these waters. No doubt the operation of Soviet fishing fleets in other parts of the world was a factor in this decision.

Besides these two types of voyage—freighters on charter to Soviet organisations and United States icebreakers on scientific cruises—there have only been very occasional naval visits by foreign ships. The best known of these was the passage along the whole length of the route of the German merchant cruiser *Komet* in 1940. The British fleet oiler *Hopemount* went from the west to Tiksi and back in 1942. Several U-boats and the cruiser *Admiral Scheer* made forays into the Kara Sea in 1942–4. There may have been other voyages of a more clandestine kind both during and after the war, but it is unlikely that there were many.

The issue as to whether the route is or is not open to the flags of all countries has always been obscured by the fact that there has been no inclination among non-Soviet shipping interests to make use of it. The Soviet government, it is now clear, is not likely to offer any legal objection. In practical terms, however, passage would be difficult if there were no Soviet co-operation because information about the ice and its expected behaviour, as well as the services of icebreakers and rescue teams, could be withheld. It is of interest, therefore, that in early 1967 the Soviet government offered to open the through route to foreign shippers. A fee, based on the net registered tonnage of the vessel, would be charged, and for this the necessary icebreaker and information services would be provided. Attention was drawn to the fact that a saving of up to thirteen days' sea time might be made on a voyage between, say, Hamburg and Yokohama. Despite an apparently successful demonstration voyage by the Soviet freighter *Novovoronezh* in 1967, the offer was not taken up. No doubt potential users feared that delays, due for instance to fog, and the additional insurance premium as well as the fee would offset the hoped-for gain in time. There is some reason to believe that the offer was later tacitly withdrawn, the presumed reason being that after the Middle East war of 1967 the Soviet Union would not wish to offend its Arab friends by opening up a viable alternative to the Suez Canal. No further mention has been made in the Soviet press on the subject of foreign use of the route. It is possible to interpret the whole incident in terms of a directive to all ministries to maximise foreign currency earnings. The Ministry of the Merchant Fleet was committed to operating an expensive route in the Arctic, and perhaps saw an opportunity to make some money on the operation and gain official approval without any further cost to itself. It is possible the offer may be renewed at some time in the future, but there is no immediate likelihood that this route may become attractive internationally.

THE FUTURE

From the fact that five, perhaps six very powerful new icebreakers are on order for the Soviet Union, one may be reasonably sure that the northern sea route has a future. Yet without that evidence there would have been some grounds for doubt. The tendency over the whole post-war period has been for access to northern Siberia from the north, via the sea route, to lose custom to access from the south, via the rivers. The Irtysh, Ob and Yenisey have been intersected by the Trans-Siberian Railway ever since its construction. The Lena was reached by a spur of that railway in 1950. The Kolyma was intersected before the Second World War by a road from Magadan on the Sea of Okhotsk, which has a notably longer ice-free season (about six months) than the northern sea route. This road was later extended to the upper Indigirka and upper Yana, and also westwards right through to the Aldan and the Lena where it joined the Amur–Yakutsk highway to reach the railway at Bolshoy Never. This major link, 3000 km long, between Magadan and the railway, was built during the war, like the Alaska highway with which it has much in common, but was not properly completed. It was turned into an all-weather road only in the 1960s. Today it still lacks bridges over several major rivers, and the central section between the Lena and the Khandyga is not yet up to all-weather standards. However, even bad roads and rivers with a six-month season offer the possibility of a better service than does a sea route with a four-month season. Thus, as the fleets of vessels operating on the rivers have grown and improved and the road system has become better, so has the southern approach gained at the expense of the northern approach. The service area of the sea route has steadily contracted and now seems to be reduced to the regions round the mouths and lower reaches of the three big rivers, extending further south only in the valleys of the smaller rivers, particularly the Khatanga, Anabar and Olenek, which are not intersected by rail or road systems.

Yet the evidence already quoted, though fragmentary, about the volume of traffic carried seems to indicate no reduction. One may infer that although the sea route is able to retain only a diminishing proportion of the traffic to and from northern Siberia, that proportion is in absolute terms remaining relatively constant, or perhaps even growing. The northern sea route is an example of building up a transport system before the purpose to be served was known in any detail. There was, as we have seen, a strategic component in the original motivation, and this is unlikely to have survived at least in the same form. It may be, therefore, that the route has failed to come up to the original, somewhat vague, expectations, but since it represents a very considerable investment of money and skills there has been reluctance to let it fall back to a more rational level. However, this hypothesis can hardly explain the orders for the new icebreakers.

There are two points which may be relevant here. The first is that if the season could be notably lengthened the route would begin to show an advantage over the down-river access because the length of season on the rivers could be extended only with great difficulty, if at all. A recent article by S. V. Slavin, one of the foremost economic planners for the Soviet north, states that the Arctic and Antarctic Research Institute is in fact working on the question of year-round navigation on the sea route, developing the idea of keeping permanently open a channel in the first-year ice (*Problemy Severa*, No. 17, 1972, p. 18). The second point is that there may be an influence from the other side of the North Pole. In 1969 and 1970 a group of oil companies interested in exploiting the discoveries at Prudhoe Bay on the north coast of Alaska performed some experiments to decide whether an Arctic sea route could be used for the export of this oil. They obtained a supertanker of 115 000 tons dwt., *Manhattan*, and had her structurally strengthened for ice navigation. The voyages she then made in the waters of the north-west passage failed for purely economic reasons to convince the companies to go ahead and build a fleet of still larger ice-strengthened tankers (as had been mooted), but *Manhattan*'s experience did show that very large ships can break much more ice than many had supposed possible. Year-round navigation seemed to be in sight, but ships like *Manhattan* and the bigger ones which might follow her would find it very difficult, probably impossible, to operate in the relatively shallow waters of the northern sea route. This direction of advance is therefore not a desirable one for the Soviet government. That government, though, would scarcely wish to sit back and see other countries acquire superior ability in a field of endeavour in which the Soviet Union has been the undisputed leader for so long. So perhaps the new icebreakers are a Soviet answer to the possibilities raised by *Manhattan*.

6

Soviet Air Transport

Air transport has evolved wholly during the Soviet period and consequently the air services of the Soviet Union owe nothing to the requirements or facilities of the pre-revolutionary period in which the broad patterns of other branches of the transport system were laid down. Since air transport serves mainly the industrial, commercial and urban populations, however, the present-day network of air routes conforms to the pattern of all transport systems operating within Soviet territory of intensive services in the western regions and a relatively sparse network in the east. Nevertheless, the flexibility of air services has enabled a much more complex pattern of routes to be evolved than in any other transport medium serving thinly populated areas.

In Soviet civil aviation it is the internal services that are most important to the development of the country, but the Soviet state airline, *Aeroflot*, also operates a worldwide network of international services. The details of these services show the influence of political and strategic as well as commercial factors, but it is another characteristic of air transport, particularly where long domestic routes are involved, that internal and external services can be closely integrated.

Another distinctive aspect of the air services as compared with other transport media over Soviet territory is the use of Soviet airspace and ground facilities by airlines and aircraft of other nationalities on a regular, scheduled basis. Such use is strictly controlled and limited to that which the Soviet authorities are prepared to authorise in exchange for their operating rights abroad, but it introduces an exotic element into the Soviet transport scene not found on the railways or inland waterways, nor, except for touring and certain trading vehicles, even on Russian roads.

The operation of civil air services has the objective of fulfilling a need for transport, whether this is already clearly realised or is latent but expected to develop when the facilities are provided. The provision of the services requires elaborate and expensive equipment in the form of aircraft and airports and supporting ground services, but no permanent way is required so that within the capabilities of the equipment and aerodromes services are infinitely variable in both direction and frequency.

It is convenient to examine first the provision of the essential aircraft and then the development of the route network, though the two cannot be completely separated. Thirdly, airports will be examined.

In the provision of aircraft it may be noted that, although all capital provision and decision-making is in the hands of the state, aircraft designers in charge of design bureaux give their own names to the products of their teams and their factories appear to concentrate production mainly on their own designs so that there is at least a superficial resemblance to the company structure of aircraft construction in capitalist countries. Furthermore, there has been marked stability in the leadership of the industry. Antonov, Ilyushin, Tupolev[1] and Yakovlev, leading names in Soviet civil aviation today, have played a major role in aircraft design for many years—some of them since the twenties.

The first regular air services in the Soviet Union began on 1 May 1922, Fokker aircraft being used by a joint Soviet–German airline on a route between Moscow and Königsberg. In March 1923 the Soviet airline *Dobrolet* was created and Moscow was linked with Nizhniy Novgorod (now Gorkiy). A number of types of foreign aircraft were used and some were built under licence in the U.S.S.R. at about this time, but several types of single-engined aircraft were designed and produced in the U.S.S.R. for civil or dual civil and military

---

[1] A. N. Tupolev died 23 Dec. 1972 at the age of 84.

purposes in the twenties. One, the AK-1, was used for the inaugural flight when the Moscow–Nizhniy Novgorod route was extended to Kazan in July 1924.[1] Skis were fitted to this machine for winter operations. In 1929 a triple-engined airliner with accommodation for nine passengers, the ANT-9, designed by A. N. Tupolev, was introduced. It was used on the Moscow–Berlin service and was also adapted to Russian winter conditions with a ski undercarriage as well as the normal wheeled type. About seventy of these aircraft were built, including a twin-engined version.[2]

In 1932 civil aviation was reorganised and the name 'Aeroflot' was adopted for the monopoly operator. In that year some 27 000 passengers and about 900 tons of freight were carried. The following year saw the opening of the main trunk route from Moscow to Irkutsk via Kazan, Sverdlovsk, Novosibirsk and Krasnoyarsk. For comparison, it may be noted that by this time British airlines were operating to South Africa and India, although as late as 1934 the journey from Croydon to Cape Town involved thirty-three separate stages in seven types of vehicle.[3] By 1936 Aeroflot had built up the network of services shown in Figure 13.

The Soviet aircraft industry had to deal with the longest domestic air routes in the world and from the beginning there was the inducement to offer air transport as a logical alternative to the ill-developed and slow surface means. Soviet designers also realised the potential economy of the large aeroplane, particularly as under the Soviet political system high load factors could be ensured by direction of travelling personnel to particular forms of transport. The propaganda value of impressively large machines also appealed to Soviet leaders, as in other spheres of mechanisation. Tupolev's eight-engined *Maxim Gorkiy* (ANT-20), designed with these differing values in mind, first flew in May 1934 but was destroyed in 1935. The six-engined ANT-20bis, developed from it, was completed in 1939 and could carry sixty-four passengers. Only one was completed but it went into service in May 1940 between Moscow and Mineralnyye Vody. The

---

[1] Stroud, 1966, 521.

[2] Ibid., 531–3.

[3] Howard and Gunston, 1972, 171.

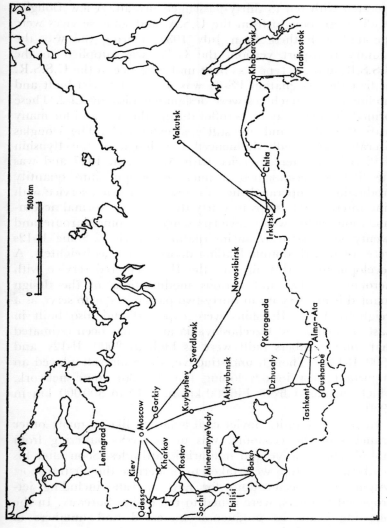

Fig. 13. Aeroflot services in 1936. *Source*: *Atlas razvitiye*

design and operating problems of such large aircraft were, however, too great for the technical knowledge of the time and most airline services in the U.S.S.R., as elsewhere, were operated with ten to thirty-passenger aircraft. A few Douglas DC-3s were bought from the U.S.A. and some services were operated with these from July 1937. During the war the military transport version, the C-47, was supplied to the U.S.S.R. and the type was built under licence in the U.S.S.R. At first the designation PS-84 was used but Soviet-built and modified U.S. machines were designated Lisunov Li-2. These formed the mainstay of Aeroflot during the war and for many years thereafter, and are still used extensively. The Douglas aircraft undoubtedly influenced the design of the Ilyushin Il-12, which entered service with Aeroflot in 1947 and was the first Soviet-designed airliner to be put into quantity production. It operated the Moscow–Vladivostok service with nine intermediate stops in thirty-three hours. Normal accommodation was for twenty-seven passengers on domestic routes and twenty-one passengers on international services. Some Il-12s were fitted with double loading doors for use as freighters. A development of this aircraft, the Il-14, entered service with Aeroflot in 1954 and various modifications of the design enabled it to carry up to thirty-two passengers or to serve as a freighter. These Ilyushins were exported and also built in East Germany and Czechoslovakia and it has been estimated that total numbers built were as high as 3000 Il-12s and 3500 Il-14s.[1] Though undistinctive, these aircraft played an important part in developing the Aeroflot route network, which increased from 146 000 km in 1940 to 300 000 km in 1950.

In the early fifties Soviet civil aviation, like so many other branches of the economy, was undoubtedly suffering from Stalin's conservative and militaristic attitudes. Soon after his death in 1953 steps were taken to modernise the air transport system, and specifications for new aircraft, including jet-propelled airliners, were issued to the design bureaux. In the following years, as production increased and numbers of passengers carried rose rapidly, the emerging aircraft types

---

[1] Stroud, 1971, 112–13.

enabled a complex hierarchy of routes to be built up with aircraft suited to one or more stages in this hierarchy. Five such stages in the hierarchy from trunk (primary) to minor local (quinary) services can be readily distinguished.

On the trunk routes, the primary network, Aeroflot entered the jet age in 1956 with the Tupolev Tu-104. Owing to the withdrawal in 1954 of the British Comets from airline service after two were destroyed in flight, the Tu-104s were the only jets in airline service anywhere in the world between 1956 and 1958. Only 50 passengers were carried in the first Tu-104s but this was increased in later versions successively to 70, 85–100 and, over certain routes, to 115. This aircraft and its derivatives still operate many of the primary and secondary Aeroflot domestic services, probably about 200 having been built.[1]

Overlapping the development of the jets came another of the most successful of Russian airliners, the Ilyushin Il-18. This four-engined transport went into service with Aeroflot in April 1959. Early models carried 80 passengers, current versions up to 110. Seating varies, as with other Soviet aircraft, according to season, because large wardrobes are necessary for stowing heavy clothing in winter and when these are taken out there may be seating for about eight extra passengers. By 1969 well over 500 had been built and Aeroflot's Il-18s had carried over 60 million passengers and were operating some 800 domestic services.[2] At first they were used for the primary services but as later aircraft have become available they have been switched to secondary trunk routes, especially to and within Central Asia, for which their long range (up to 3700 km with maximum payload) suits them.

Although they had not produced any modern commercial aircraft larger than the Il-14 up to the mid-fifties, by 1961 Russian designers were challenging the world with their large civil aircraft. In that year the Tupolev Tu-114 went into service. This was the largest and fastest propeller-turbine transport, capable of carrying up to 220 passengers, about twice the number that contemporary airliners could carry,

---

[1] Stroud, 1968, 209.
[2] Taylor (ed.), 1970, 489.

over a range of about 10 000 km, cruising at 770 km/h (nearly 480 m.p.h.) The Tu-114 flew the primary Moscow–Khabarovsk route, 6200 km (3820 miles) in eight and a quarter hours. It also pioneered Aeroflot's transatlantic services to Havana in 1963 and Montreal in 1966 and was used on the Delhi, Accra and Tokyo services. At least thirty of these aircraft were built and many are still in service, based at Moscow. On most of the long-distance international services, such as those to Montreal and New York, the Tu-114 has been replaced by the Ilyushin Il-62 four-jet transport with accommodation for 168 passengers in the standard version.

The problem of the most efficient and most economical ways of using short, unpaved and rough airfields by medium-sized aircraft received much attention from Soviet designers. Among aircraft designed with these needs in mind was the Tu-124, a smaller version of the Tu-104, to replace the old Il-14 but with seating for fifty-six passengers. This type went into service more than two and a half years ahead of the first western short-haul turbofan transport.[1] Two years later (1964) the larger twin-jet Tu-134, capable of carrying up to seventy-six passengers, appeared with many features to fit the needs of secondary routes, short stages and quick turnround. An entirely new design, the Tu-154, a tri-jet capable of carrying up to 158 passengers was announced in 1966. This type represented another major step forward in the medium-range aircraft. The Tu-154 has a maximum range of over 6000 km and is capable of making regular use of airfields with surfaces of packed earth and gravel.[2] This type will eventually replace the Tu-104 and Il-18 on many routes which may be classed generally as secondary.

FEEDERLINE AIRCRAFT

The feederline (tertiary and quaternary) services remained without new equipment until 1962. More than 200 federal routes were then being operated by elderly piston-engined

---

[1] Stroud, 1966, 573.
[2] Taylor (ed.), 1970, 518.

Il-14s, Il-12s and Li-2s.[1] In 1962, however, the An-24, with seats for forty-four passengers, began to arrive in service on the Moscow–Voronezh and Saratov routes. Like its British and Dutch counterparts, the Handley Page Herald and Fokker Friendship, it can operate from small airports with paved or unpaved runways. Following a decision to provide fast jet services from a large number of these airfields, the Yakovlev design bureau produced the Yak-40, the first jet transport intended for grass airfields and unpaved runways of 1000 m or less.[2] On short (quaternary) routes the Yak-40 tri-jet is replacing aircraft such as the Li-2 and Il-14, and is expected eventually to operate over several thousand route stages.

Serving some of the quaternary routes, and also quinary ones to very small communities, the humble An-2 single-engined biplane has been one of the most successful types of aircraft built in the Soviet Union. It carries only seven to ten passengers but as a whole the type had carried 100 million passengers by 1967, and in 1970 was operating more than 2000 local route stages and handling almost 40 per cent of Aeroflot's passengers on local services.[3] These simple statistics emphasise the high proportion of Aeroflot's traffic formed by relatively short journeys from grass and snow-covered airfields. The An-2M is one of the world's most successful agricultural aircraft, used all over the Soviet Union and in many other countries for crop spraying and fertiliser application. It is also used for forestry work, fire-fighting, medical and ambulance services and a host of other jobs, all of which are the responsibility of the Aeroflot monopoly. The Czech Let L-410 is intended partially to replace the An-2. Standard seating in this adaptable twin-engined aircraft is for 15–19 passengers but it will have many alternative roles including freighting and ambulance work.

The problem of operating from earth and grass airfields in summer and snow in winter with heavy loads was largely solved by 1959 with the Antonov An-10 carrying 80–100 passengers and its derivative, the An-12, the mainstay of Aeroflot's freight services. These machines can operate from

---
[1] Stroud, 1968, 78.
[2] Ibid., 270; Taylor (ed.), 1970, 523.
[3] Taylor (ed.), 1970, 479.

prepared airfields or virgin snow and have heated, brake-equipped skis for this purpose.

A new cargo-carrying giant appeared as an unexpected visitor at the Paris Air Show in 1965. This was the Antonov An-22. It arrived carrying three buses and general cargo, and proved itself capable of lifting 100 000 kg. Among the many tasks which the type has since performed has been the delivery of mobile power stations and drilling rigs to the oil fields around Tyumen. The airlifting of rigs and other heavy equipment is also a task for the giant helicopters or 'flying cranes', a Russian speciality. Another new cargo aircraft, the Il-76, suitable for operation from rough airstrips in the north and Far East, appeared in 1971.

In contrast, Soviet preparations for supersonic air transport have proceeded simultaneously with the development of the Anglo-French Concorde. It has been officially stated that the Tu-144, which is in series production, should enter service with Aeroflot between Moscow and New York and Moscow and Tokyo in 1975 and co-operation with British Airways and Air France on these sectors is envisaged.[1]

ORGANISATION OF AEROFLOT

The departmental and regional administration of Aeroflot has been changed from time to time. By 1964 regional managements had evolved into eighteen Territorial Directorates, nine Aviation Groups and two special organisations, for international services (TUMVL) and polar operations respectively. By 1968 all these had become Directorates. In 1973 a new Directorate was formed, based at Arkhangelsk, and the thirty and their headquarters then were:

Arkhangelsk	Arkhangelsk	Kazakh	Alma-Ata
Armenia	Yerevan	Kirgiz	Frunze
Azerbaydzhan	Baku	Komi	Syktyvkar
Belorussia	Minsk	Krasnoyarsk	Krasnoyarsk
Eastern Siberia	Irkutsk	Latvia	Riga
Estonia	Tallin	Leningrad	Leningrad
Far East	Khabarovsk	Lithuania	Vilnyus
Georgia	Tbilisi	Magadan	Magadan

[1] *Flight*, No. 3301, 15 June 1972, 839.

Moldavia	Kishinev	Volga	Kuybyshev
Moscow	Moscow–Bykovo	West Siberia	Novosibirsk
Moscow Transport Directorate	Moscow–Vnukovo	Yakut	Yakutsk
North Caucasus	Rostov	Central Regions and Arctic (UGATsRiA)	Moscow–Bykovo
Tadzhik	Dushanbe		
Turkmen	Ashkhabad	International (TsUMVS formerly TUMVL)	Moscow–Sheremetevo
Ukraine	Kiev		
Ural	Sverdlovsk		
Uzbek	Tashkent		

The organisation appears to have been based on regional autonomy with each directorate having its own headquarters, aircraft and maintenance staff and being responsible for operations within its own area and specified federal operations, either solely or in pool.[1]

The Aeroflot network totals nearly 800 000 km. Of this total international routes account for 225 000 km, serving sixty-three countries in Europe, Africa, Asia, North America and Cuba (Fig. 14). It is intended to extend the network to Australia and South America as well as to other parts of Africa during the period 1974–7. Most of the international routes are operated by TsUMVS, based at Sheremetevo Airport, Moscow. There are also international services from a number of other cities, such as from Kiev to Berlin, Budapest, Bratislava, Prague, Vienna, Warsaw and Zurich, and from Leningrad to Stockholm, Copenhagen, Amsterdam, Paris and London. Some of these services are operated by the regional Directorates, notably the Ukrainian.

The amount of traffic carried by Aeroflot continues to increase rapidly. In total scheduled traffic in 1971 the U.S.S.R. was second only to the U.S.A. with 9915 million ton-km compared with the U.S.A.'s 27 280 million. The Soviet total was thus about two-fifths of the American. For comparison, it may be noted that U.K. airlines are third in this league of total scheduled ton-km performed, but with only about one-quarter of the Soviet traffic.

A breakdown of these figures reveals that the number of passengers carried by Aeroflot was 71 million in 1970 and 78 million in 1971, the latter figure representing one in five of all passengers carried on scheduled flights throughout the world.

---

[1] Stroud, 1968, 32.

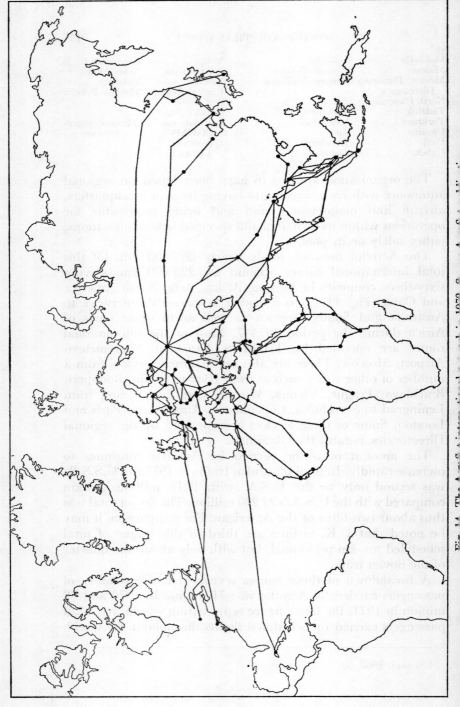

Passenger-kilometres operated were 78 200 million in 1970 and 88 800 million in 1971, a little less than one-fifth of the world total, the proportion being lowered by the effect of the very long overseas and round-the-world routes of many other airlines. In freight carriage the Soviet totals in the two years were, for 1970, 1877 million ton-km, and 1971, 1982 million ton-km, about one-sixth of world totals.[1] No comparative figures are available for non-scheduled traffic.

DOMESTIC SERVICES

Moscow is the hub of the air service network and has direct connections with over 120 Soviet cities. Whether direct or not, several flights are available daily to most cities in the U.S.S.R. from the capital (Fig. 15). The longest route, the trans-Siberian, connects Moscow with Novosibirsk in 5 hours, Irkutsk in 6 hours 50 minutes, Khabarovsk in 7 hours 35 minutes and Vladivostok in 11 hours 25 minutes, compared with 2 days, $3\frac{1}{2}$ days, 6 days and 7 days respectively by rail, illustrating the time-saving that can be achieved by air travel. Non-stop flights have been available over the 6400-km route to Khabarovsk from Moscow since the introduction of the Tu-114 in 1961. Long-distance direct connections are available between many Soviet cities without the necessity of travelling via Moscow, some of these connections being shown in Fig. 15. Thus, in the summer of 1972, Novosibirsk could be reached in 6 hours 10 minutes from Tbilisi and 5 hours 55 minutes from Kiev. The departure schedule from Kiev showed over fifty destinations, excluding local services, with Barnaul, Tyumen and Tselinograd illustrative of the distant cities featured. From Tbilisi, forty-six services were shown, including Novosibirsk, Sverdlovsk and Chelyabinsk.

The advantage of the aeroplane for the long journey is

---

[1] More correctly, among member-countries of the International Civil Aviation Organisation. The chief non-contracting country is China. Figures from *Icao Bulletin*, May 1972 (*Flight*, vol. 101, No. 3300, 8 June 1972, 823–4). *Flight*, 28 June 1973, 976, *Tran. i svyaz'*, 1972, 208.

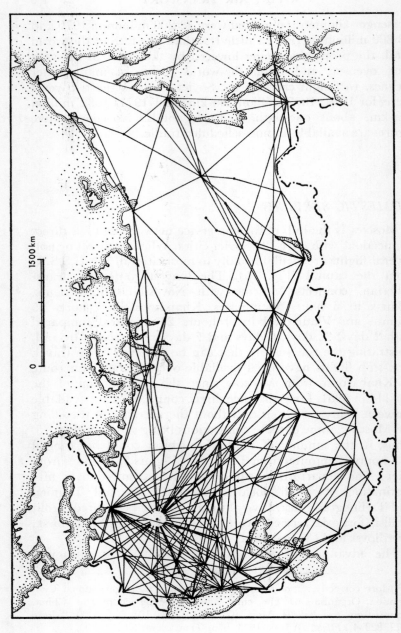

Fig. 15. Internal air routes of major interregional significance operating in 1972. *Source*: Aeroflot publications, *Atlas razvitie*

demonstrated by the average length of passenger flight—1001 km in 1966 when the average rail journey was 90 km. Even in 1928 the average flight distance was 413 km and it had risen to 1213 km in 1955, but remained in the 700–800 km class for a number of years thereafter, probably because of the opening of new short-haul routes. The distance factor explains Aeroflot's contribution of no less than 10 per cent to the total Soviet passenger movement, expressed in passenger-kilometres. On the other hand, only about 0·2 per cent of all journeys originating are by air, but the grand total of over 21 million journeys by all carriers includes suburban traffic and the true importance of the airways is probably best gathered by comparing air departures with originating journeys by train, excluding suburban traffic.

In 1965, when the total passengers carried by Aeroflot in the U.S.S.R. numbered 42 070 000, this equalled 16·5 per cent of all non-suburban rail passengers. Of this number 25 661 000 originated in the R.S.F.S.R., 5 413 000 in the Ukraine and 3 046 000 in Kazakhstan. The regional variation in the importance to the passenger is emphasised by the air/rail comparison. For every 100 passengers setting off by non-suburban train, Aeroflot recorded 10–17 passengers in the Ukraine, Belorussia, each of the Baltic republics and Georgia, 19 in Tadzhikistan, 30 in Azerbaydzhan and over 40 in Turkmenistan, Uzbekistan and Armenia.[1]

Since 1965 Aeroflot has increased its performance in all respects with larger and faster aircraft providing more frequent services and the extension of the network, and it is official policy to encourage the use of the airlines for medium and long-distance travel.

Yet it is the local services that are the most distinctive and perhaps the most interesting. As well as the interregional services there is a network of short-haul routes around each regional centre. These could not be shown on a map with a scale similar to that of Figure 15 but a regional example is given in Figure 16, and, by comparing this with the appropriate zone in Figure 15, an impression can be gained of the intricacy of the complete network.

---

[1] *Transport i svyaz'*, 1967, 108–9 and 220.

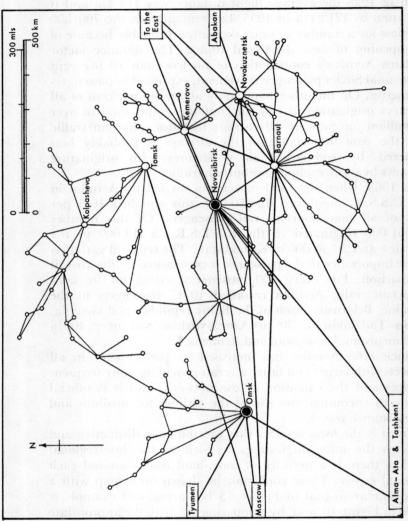

Fig. 16. The Aeroflot network in part of western Siberia. The map gives an indication of the hierarchy of services from transcontinental to local. *Source*: Aeroflot publications.

The total network coverage is very impressive, but it must be remembered that there is no private flying in the U.S.S.R. This does not mean merely that there is no operation privately of light aircraft for sporting and recreation purposes but also there are no private business aircraft. The latter are an important part of civil aviation in North America and western Europe and remove from the airlines a regular source of revenue which is available to Aeroflot. It is, of course, possible for the Soviet government to direct virtually all its citizens to use the means of travel which it prefers if it wishes so to do. The limitations placed on private motoring, the climatic rigours and the absence of good roads over large areas in any case restrict the alternatives, while the advantages of flying over any form of surface travel over long distances have already been remarked.

AIRPORTS

Soviet airports, together with navigational aids and maintenance and training establishments, are operated by Aeroflot, under the control of the Ministry of Civil Aviation.

Several systems of classification of airports appear to be used in the Soviet Union. Two of these express in general terms the functions of the airports, the one emphasising the comparative importance of an aerodrome to the airlines, and the other the types of services provided.

Each of these two classifications recognises three groups. The first couples location with importance to the airline.

1. Fixed (permanent) bases (*bazovyye*) of one or several of the subdivisions of the air services. Such would be the airports at the places named as headquarters of Aeroflot Directorates, but the term would probably apply to other main airports strategically located.

2. Terminal (*konechnye*) airports, where aircraft complete their scheduled flights and are serviced. Some of these are *bazovyye*, but the term would extend to all important airports where flights terminate and relevant facilities are provided. Whether the term applies to airports which are at the end of a flight but have few or no maintenance facilities is not clear.

3. Intermediate (*promezhutochnye*) airports, where short halts are made for traffic or technical reasons.[1] Main bases may also, of course, be used in this manner by through services.

Secondly, the types of services provided and the role in the network of communications are expressed in the complementary classification into airports of (*a*) international, (*b*) union and (*c*) local significance.[2] More important is the technical classification of facilities according to which the traffic capacity of an airport is determined. For this purpose Soviet airports have been divided into five classes and unclassified aerodromes.[3]

Class I and Class II airports are those that handle a sufficient volume and complexity of traffic for there to be an establishment of numerous specialised departments to deal with all the functions of the airport— commercial, operational, security, passenger services, freight handling, motor transport, catering, fuel and lubricants supply, etc. Airports of Classes III and IV justify a less specialised departmental organisation in accordance with less extensive and less intensive utilisation. In Class V airports there is no need for this departmental organisation and all commercial and operational functions can be supervised directly by the airport manager. The layout of buildings is necessarily more complex in the higher class airports. Classes I–III airports justify special accommodation for personnel for servicing of aircraft and machinery grouped with the technical facilities of the maintenance hangars—in Soviet jargon the *Aviatsionno-tekhnicheskiy baz* (ATB)—whereas in Classes IV and V airports technical staff are less numerous and can be housed in the main terminal buildings.

The length of runways needed for each class of airport is not specified in this published synopsis of the classification, although the scale of provision of runways is fundamental to the scope of operations that can be carried on at any airport. Nor is the present writer aware of any published data stating the classes ascribed to individual airports. Such technical information as is available on Soviet airports is limited to that required by the foreign airlines permitted to operate over Soviet air space. This, however, includes data on the airports of the

---

[1] Mukhordykh, 1971, 95–96.
[2] Ibid., 96.
[3] Ibid., 100.

largest cities (Moscow, Leningrad, Kiev, Tashkent, etc.) and some of those which are important as diversionary airports for emergency use, and there are sufficient examples to indicate the variation of airport size and facilities found on main routes.

We will consider first the data available on actual runway lengths and widths. Table 24 gives examples of a range of Soviet airports.

TABLE 24

U.S.S.R. INTERNATIONAL AIRPORT RUNWAYS

(metres)

	Paved Runways		Grass Runways	
	Main	Secondary	Main	Secondary
Moscow				
Sheremetevo	3500 × 80	—	—	—
Moscow Vnukovo	3050 × 60	3000 × 80	1500 × 40	1500 × 60
Leningrad	3400 × 70	2500 × 60	1700 × 50	—
Kiev Borispol	3500 × 63	2500 × 60	3850 × 100	1200 × 60
Novosibirsk				
Tolmachevo	3600 × 80	1000 × 80	2800 × 100	—
Ryazan Diagilevo	3000 × 70	—	3000 × 100	—
Riga (Central)	1700 × 50	—	1500 × 100	550 × 70

Source: International Aeradio.

Classification of the airport in terms of operating capabilities depends primarily on the paved runways. It is generally considered that the main runway of a full international airport should extend to at least 3500 m. Furthermore, the actual length of any runway is subject to modification (in assessment of its operational value) by factors which take into account the altitude of the airport, the temperature and humidity regimes and other conditions which affect the take-off performance of aircraft.[1] Of the examples given, Moscow Sheremetevo, Leningrad, Kiev and Novosibirsk are not greatly affected by these factors and the runways may be rated as qualifying for the Class I designation. They are the best found in the U.S.S.R., at least on civil aerodromes, and though not of

---

[1] Annex 14, 1971, 17–18.

overgenerous length they are adequate for the aircraft currently in use operating at maximum weights, as is desirable for the long Siberian and Central Asian sectors. Surfaces are of concrete and maintenance is generally at a high level with suitable machinery for clearing the snow, ice and slush that are frequently present in winter.

The capacity of an airport is increased and the risk of crosswinds or accidents putting it out of action is decreased by the provision of additional runways. It could be argued that any Class I airport should have a second runway, though Sheremetevo has only one. Twin, parallel runways, as at Kiev Borispol, permit simultaneous movements such as taking off and landing on the adjacent runways. Divergent runways, such as at Leningrad, facilitate high-wind operations. Novosibirsk has an interesting pattern with one main runway and two short, angled runways. These, 1000 m long, can be used by medium-sized aircraft with STOL (short take-off and landing) characteristics. The Yak-40 feederline twin-jet needs only 750 m for take-off. It can also operate from grass so that with this aircraft even jet operations are feasible on the subsidiary strips.

All operating facilities for safe and frequent movements of large, fast passenger aircraft by day and night are provided at these major airports. Terminal facilities are comprehensive, including, at Moscow, Leningrad, Kiev and Tashkent, full customs, and health and passport-control services. But, even at Moscow Sheremetevo, which handles the bulk of the international traffic, passenger services are limited to those essential for international movement and the basic requirements for the comfort of passengers—such as buffet and cloakroom facilities, a bank, post office and 'Beriozka' shops selling Soviet goods for foreign currency, and souvenir stalls. There is no multiplicity of shops such as is found in the leading airports of western countries such as London Heathrow, Amsterdam, New York, etc.

Facilities at some Soviet airports are restricted by the available space in the existing terminal buildings, as at Sheremetevo, which are more in keeping with those normally provided at western capitals in the fifties than the present time. The concourse available for passengers awaiting flights prior to passing customs and passport control is of very limited size. There is a more generous lounge and restaurant beyond these checks but passengers may not proceed until the flight is called and so the use of these facilities is curtailed. Final departure is from a single tower at the end of an approach corridor. The tower has only three gates, which simplifies control with a limited number of flights, but such an arrangement would soon prove inadequate if movements increased substantially or jumbo-size aircraft were used. For arriving passengers also, facilities are limited with few customs points and no segregation of returning Soviet citizens, who are liable to have their baggage searched thoroughly for forbidden or dutiable goods, and foreign tourists who are usually passed with only a cursory inspection.

Intourist officers assist in controlling the departure of foreign visitors from airports and frequently have entirely separate waiting rooms for foreigners, even in relatively small airports having only domestic services. By western standards such arrangements are space-wasting, though they do ensure the waiting tourist a degree of quiet comfort rarely met elsewhere.

While the runways of the first five airports shown in Table 24 would justify them being described as Class I airports, Ryazan is more restricted. It is, however, capable of receiving large aircraft and is nominated as a diversionary airport for foreign transcontinental flights. By international standards it is of second rather than first-class technical level, but its suitability for such aircraft as the Tu-104 and Tu-134, which can operate sectors of 2000–3000 km, has ensured the provision of long-distance direct flights to a wide variety of destinations. Many cities have a similar quality of services, both in direct links and in complexity of the network into which they fit, based on airports of the size-group with runways of 2500–3000 + m, other examples being Irkutsk, a major airport on the trans-Siberian route with 2750 m of runway, and Tashkent. There is a runway of 3500 m at Tashkent but the temperature factor reduces its effectiveness to the equivalent of less than 3000 m, without, however, bringing it below the safe requirement of an Il-62 or a DC-8. Samarkand, with a 2400 m runway, similarly affected by temperature, must be considered to be in a class lower in the hierarchy but still has direct services from Moscow by the Il-18 turboprop aircraft. Riga, with a longest runway of only 1700 m, is in yet a lower class. This runway is inadequate for normal service with heavily loaded jets such as the Il-62 and DC-8, but with Il-18s, Tu-134s and other aircraft, Riga is provided with direct services to the Black Sea coast and Caucasus and the services will improve with newer aircraft such as the Tu-154 which has greater capacity and longer range, but a very moderate runway requirement.

With the STOL aircraft now developed the U.S.S.R. is in a position to give excellent air services to towns and districts with quite small airports. Not only can they have modern aircraft such as the Tu-154, Tu-124, or at least the Yak-40, to serve them, but with the long range of which the Tu-154 is capable, direct services can be provided between points with substantial traffic such as industrial towns in Siberia and the Black Sea coast, without interchange. This, in turn, reduces congestion at Moscow and other main airports.

The opportunities of STOL aircraft substantially reduce the capital cost of providing facilities for the scattered population of the Union. Although the air services are run as a state monopoly fares are kept low and state subsidies cannot be unlimited. Hence, the required services must be provided at minimum cost consistent with safety. The less elaborate that airports have to be for the level of services and capacities determined, the more economic the services will be. The Soviet Union has been pursuing a policy of transferring almost all long-distance travel and much of that over quite short distances to the airways, and the present technical developments should be helping them with this aim.

It has also long been Soviet policy to provide air services for small and remote communities through aircraft that can operate from airstrips of the minimum degree of refinement. This has been achieved mainly through the services of the An-2 biplane, now to be gradually replaced by the L-410, a twin-engined STOL monoplane, and, on suitable routes, by the Yak-40.

The STOL designs, it may also be remarked, reduce demand for land for airports, and, although the Soviet Union is the largest country in the world it can afford no unnecessary losses of agricultural land. Furthermore, STOL aircraft gain height rapidly and reduce the impact of noise on surrounding areas, so this type of design is doubly desirable in environmental terms.

BIBLIOGRAPHY

ABC World Airways Guide, Dunstable, monthly
Airports International, London, monthly
Annex 14 to the Convention of International Civil Aviation, *Aerodromes*, 6th edition, Montreal, I.C.A.O., 1971
Flight International, London, weekly
Grazhdanskaya aviatsiya, Moscow, monthly
Howard, F. and Gunston, W., *The conquest of the air*, London, Elek, 1971
International Aeradio Ltd., London, maps and charts
Kish, G., Soviet air transport, *Geographical Review*, vol. 48, 1958
Mukhordykh, E. V. in Gromov, N. N. and others, *Ekonomika vozdushnogo transporta*, Moscow, Transport, 1971
Narodnoye khozyaystvo SSSR v 1972g., Moscow, 1973
Stroud, J. *European transport aircraft since 1910*, London, Putnam, 1966
Stroud, J. *Soviet transport aircraft since 1945*, London, Putnam, 1968
Stroud, J. *The world's airliners*, London, The Bodley Head, 1971
Taylor, J. W. R., *Jane's all the world's aircraft, 1970–71*, London, Sampson Low, 1970
Taylor, J. W. R., *Jane's all the world's aircraft, 1972–73*, London, Sampson Low, 1972
Transport i svyaz' SSSR, Moscow, Statistika, 1967, 1972

7

Conclusions

The chapters of this book have spanned over one hundred years of development of Russian transport. The studies of railway development and its effects in the specific cases of grain movement in European Russia and cotton movement in Turkestan in the nineteenth and early twentieth centuries provide two detailed case histories to advance our knowledge of the effects of transport development in tsarist Russia. In the following chapters there is a wide-ranging survey of rail, sea and air transport in the contemporary Soviet state. The contrast between the political and social fabric of the U.S.S.R. today and that of pre-revolutionary Russia is rivalled by the contrast in transport methods and facilities. The first two studies are concerned with the carriage of agricultural commodities—by far the dominant traffic of the period—and the replacement of cart and caravan by canal and railway. The railway swiftly attained a position of dominance in Russian transport, a position which it still holds today, as shown in the third chapter.

In spite of this dominant position investment in the railway network has never been sufficient to ensure a system capable of handling the demands placed upon it without severe strain. Prior to the Revolution foreign capital was important, but during the Soviet period this has been excluded. Transport as a whole has been traditionally afforded a place in Soviet investment inferior to that of extractive and manufacturing industry, and within the transport sector new demands have arisen. The motor-vehicle revolution, although curtailed and delayed in its effect in the Soviet Union, has led to increasing demand for investment in roads and vehicle plants. The development of overseas trade, together with the advantages of earning foreign currency at the same time as increasing strategic independence, has favoured the growth of the merchant marine. Perhaps the most dramatic transport development in recent years has, however, been the upsurge in air transport, and in this field the Soviet Union has been among the foremost in the world in the creation of a complex internal network. Soviet aircraft also provide links with many foreign countries, though there is a strong political influence in the choice of countries served and Aeroflot still remains somewhat apart from the other leading international airlines. While uncommitted countries have not shown much interest in purchasing Soviet aircraft all Aeroflot services are supplied with Russian aircraft except for a very small number of Czech machines. Thus, the Soviet Union has bridged the technology gap since the early thirties when the majority of aircraft in use in the U.S.S.R. were foreign-built.

Air transport has largely taken care of the growing long-distance movement of people while short-haul traffic, both passenger and freight, is now being catered for to a considerable extent by motor vehicles, the output of which is being rapidly increased. However, the combination of the great territorial extent of the U.S.S.R. and harsh climatic conditions has greatly restricted the role of motor transport in total movement-effort and, whereas air transport is able to deal with long-distance passenger movement, it cannot as yet offer more than a marginal contribution to the freight-movement problem. Hence, railway development has continued since the Second World War in a manner unparalleled in any other relatively developed country.

Soviet statistics, although lacking in much of the detail available for other developed countries and subjected to considerable criticism by foreign observers on grounds of inaccuracy, do provide sufficient material for the analysis of transport developments to a much greater extent than has yet been undertaken. A study in depth is not practicable in this brief concluding chapter but some salient features of the changes in the balance of transport media since the Revolution and, more particularly, in recent years, will be examined.

Although the figures quoted, especially for the earlier years, must be treated as approximations, and complete comparability over the years cannot be relied on, the magnitude of increases in traffic and of changes in the media employed make possible the identification of trends which are quite clear.

The latest figures available at the time of writing[1] show the total quantities of freight moved by each of the main media and these are given in Table 25, together with comparable figures for 1970 and earlier decennial periods.

TABLE 25

FREIGHT TRANSPORTED IN THE U.S.S.R. AND
PRE-REVOLUTIONARY RUSSIA BY COMMON CARRIERS

(million tons)

Year	Rail	Sea	Inland waterways	Pipeline	Road	Air
1913*	157·6	15·1	35·1	0·4	10·0	—
1928	156·2	8·0	18·3	1·1	20·0	0·0
1940	592·6	31·2	73·1	7·9	858·6	0·06
1950	834·3	33·7	91·8	15·3	1,859·2	0·16
1960	1,884·9	75·9	210·3	129·9	8,492·7	0·70
1970	2,896·0	161·9	357·8	339·9	14,622·8	1·8
1971	3,048·8	170·9	380·7	352·6	15,760·0	2·0

* Area within contemporary U.S.S.R. frontiers.
Source: *Transport i svyaz' SSSR*, 1972, 21.

These figures show that the total weight of goods moved by rail in 1971 was roughly nineteen times that of 1913, while sea and inland waterway transport were each about eleven times greater. In 1971 rail transport accounted for about eight times more movement by weight than the inland waterways handled. Pipeline transport of oil and gas has grown from less than one-half of a million tons in 1913 to over 350 million tons in 1971, comparable to the inland waterway figure. The greatest growth has, however, been in road transport, which now accounts for five times as much weight lifted as loaded on the railways. Road movement is, however, mostly over short

---

[1] *Transport i svyaz' SSSR*, Moscow, 1972.

CONCLUSIONS

distances and the balance of importance in terms of total transport effort is redressed when movement in terms of ton-kilometres is considered.

Movement in ton-kilometres is shown in Table 26 and the figures for the years shown in the table and intervening periods are graphed in Figure 17 on a semi-logarithmic scale to show comparable rates of increase of traffic movement.

TABLE 26

TOTAL FREIGHT MOVEMENT IN THE U.S.S.R. AND PRE-REVOLUTIONARY RUSSIA BY COMMON CARRIERS

(milliard ton-km)

Year	All kinds of transport	Rail	Sea	Inland waterways	Pipeline	Road	Air
1913*	126·0	76·4	20·3	28·9	0·3	0·1	—
1928	119·5	93·4	9·3	15·9	0·7	0·2	0·0
1940	487·6	415·0	23·8	36·1	3·8	8·9	0·02
1950	713·3	602·3	39·7	46·2	4·9	20·1	0·14
1960	1885·7	1504·3	131·5	99·6	51·2	98·5	0·56
1970	3829·2	2494·7	656·1	174·0	281·7	220·8	1·88
1971	4085·5	2637·3	696·0	183·7	328·5	238·0	1·98

* Area within contemporary U.S.S.R. frontiers.
Source: *Transport i svyaz' SSSR*, 1972, 17.

Once the factor of distance is introduced the justification for describing the railways as dominant in Soviet freight transport is clearly seen. Railway ton-kilometres performed were more than ten times the road transport figure. Similarly, these figures illustrate the post-war demands on the railways, with total movement of rail freight in 1971 more than four times the 1950 figure and 75 per cent more than in 1960. The increase in tonnage loaded over the period has been slightly less, the average haul having increased from 722 km in 1950 to 798 km in 1960 and 865 km in 1971.[1] The contribution of the railways to the gross total freight movement has, of course, declined with the development of the other media. The top

[1] *Tran. i svyaz'*, 1972, 95.

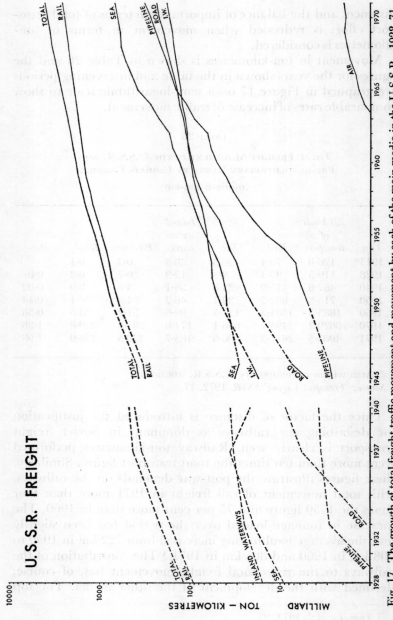

Fig. 17. The growth of total freight traffic movement and movement by each of the main media in the U.S.S.R., 1928–71.

line in Figure 17 shows the remarkably steady growth of total freight movement, falling off slightly since 1968 but still rising strongly, with the contribution of the railways rising on an almost parallel trend until 1960, then at a slightly lower rate to 1968, after which the two lines are again parallel. The logarithmic scale overstates visually the contribution of the railways, as it does even more markedly that of the lesser media, but stresses the rate of increase of each medium, while the tables facilitate the correct interpretation of the relative shares. Thus, the railways accounted for about 65 per cent of total movement in 1971 compared with nearly 80 per cent in 1960, but the increase in burden of the railways over this period has been no less than 1133 milliard ton-km—which was nearly as much as all other media added together moved in 1971.

During this same period the other media have increased their combined performance by 1067 milliard ton-km, slightly less than the railways, but representing a rapid increase over their previous figures, as illustrated graphically in Figure 17.

Although the railways had far surpassed canal and river transport in total traffic before the Revolution, the inland waterways were still comfortably in second place, loading about one-quarter as much freight as the railways, with longer average hauls raising their share in terms of ton-kilometres performed. The inland waterways lost more of their traffic during the post-revolutionary years and the subsequent build-up was slower, the figure of 100 milliard ton-km being reached only in 1960 when the railway figure was over 1500 milliard ton-km. Subsequent increases have been steady but relatively unspectacular so that the contribution of the inland waterways is now about one-eighth of that of the railways in terms of tonnage loaded but only one-fourteenth in terms of ton-kilometres performed.

In 1913 and 1917 the only other important form of transport, other than cartage, was by sea. Like river transport, shipping movements were particularly curtailed after the war and the Revolution and not until 1932 were the pre-war totals approached. Growth remained moderate, exceeding fifty milliard ton-km only in 1953, but with the extension of Soviet world trade and, in particular, relations with the Third World after the death of Stalin, the merchant marine was enlarged as described in Chapter 4 and in the five-year period 1953-8 the ton-kilometres of shipping transport more than doubled. They doubled again in the next five years and again in the next, to 586·8 milliard ton-km in 1968,[1] increasing further as shown in Table 26. Increased length of haul contributed to these impressive figures, the average rising from 1058 km in 1953 to 1503 in 1958, 2418 in 1963 and 4006 km in 1968.[2] These figures reflect the rapid increase in foreign trade with tonnages loaded on voyages abroad rising much more rapidly than tonnages moved internally.[3]

The role of railways, inland waterways and shipping has been substantially usurped in the movement of oil and natural gas within the U.S.S.R. and to other eastern and central European countries by pipelines. The graphs and tables show the rapid increase in movement by pipeline since 1953, originating tonnage being, in 1971, similar to that of all goods dispatched by inland waterway, with a longer average distance of movement resulting in ton-kilometres performed being 80 per cent higher than the inland waterway figure.

Road transport showed a higher rate of increase than all other forms of transport in most years between 1932 and 1953 (when pipeline transport increased abruptly as new pipelines came into use) but the total movement involved remained modest. Since 1953-4, the rate of growth in ton-kilometres moved has been less than that recorded by sea and pipeline, but the absolute increase in tonnages loaded on to road vehicles has risen to about four times greater than that loaded on all other transport media. Shortness of haul, however, keeps the ton-kilometres figure below rail, sea and pipeline figures.

Finally, air transport makes only a marginal contribution to freight movement and, after a rapid rise in ton-kilometres achieved between 1963 and 1969, there was a slight drop in 1970 and a recovery in 1971 only to the 1969 figure. Stability in the length of haul has resulted in the increase in recent years being mainly in higher tonnages loaded, these rising from one million tons in 1963 to 1·8 million in 1969 and in 1970, 2·0 million in 1971 and 2·1 million tons in 1972.[4]

---

[1] *Tran. i svyaz'*, 1972, 17.   [2] Ibid., 139.
[3] Ibid., 144-6.   [4] *Grazhdanskaya aviatsiya*, No. 5, 1973.

CONCLUSIONS

As may be expected, when we turn to passenger transport we find substantial differences in the balance of the contribution of the various media, but the dominance of road transport in terms of number of originating journeys remains, while in passenger-kilometres recorded the railways again lead, but with road traffic fast closing the gap and the airlines still far behind but catching up on both roads and railways. The salient figures appear in Tables 27 and 28 for the same dates as those already given for freight movement.

TABLE 27

Passengers Carried in the U.S.S.R. and Pre-revolutionary Russia by Common Carriers

(million persons)

Year	Rail	Sea	Inland waterways	Road (public service buses)	Air
1913*	248·5	3·7	11·5	—	—
1928	291·1	1·2	17·8	660	0·007
1940	1,343·5	9·6	73·0	590	0·4
1950	1,163·8	7·9	53·6	1,053	1·5
1960	1,949·7	26·7	118·6	11,316	16·0
1970	2,930·4	38·5	145·2	26,365	71·4
1971	3,053·4	38·4	145·7	27,675	78·1

* Area within contemporary U.S.S.R. frontiers.
Source: *Transport i svyaz' SSSR*, 1972, 24.

Urban and suburban transport accounts for a large part of the increase on the railways and most of the road passenger movement. As more long-distance traffic is transferred to the airways the role of the railways in suburban transport has become relatively, as well as absolutely, more important with passengers classed as suburban rising from 82 per cent of the total in 1950 to 88 per cent in 1960 and 89 per cent in 1970 (2615·9 million originating passengers out of the total of 2930·4 million). Comparable figures of passenger-kilometres for suburban services in relation to the total network were, however, only 24 per cent in 1950 and 1960 and 27 per cent in 1970. The overall average length of railway journeys continued to fluctuate around 90 km while the average suburban journey increased from 24 km in 1960 to 27 km in 1970 as suburbs extended outward from city centres.[1]

Compared with other relatively advanced countries, the U.S.S.R. was slow to develop road transport for passengers, so that even in terms of originating journeys the road services did not carry as many passengers as the railways until 1951. By 1960 they carried six times as many and in 1971 eight times as many. In terms of passenger-kilometres the road services were unimportant compared with the railways in 1950 but reached one-half of the railway figure in 1964 and over three-quarters in 1971, with their vast number of short journeys.

TABLE 28

TOTAL PASSENGER MOVEMENT IN THE U.S.S.R. AND PRE-REVOLUTIONARY RUSSIA, BY COMMON CARRIERS

(milliard passenger-km)

Year	All kinds of transport	Rail	Sea	Inland waterways	Road (bus)	Air
1913*	32.7	30.3	1.0	1.4	—	—
1928	27.1	24.5	0.3	2.1	0.2	—
1940	106.3	98.0	0.9	3.8	3.4	0.2
1950	98.3	88.0	1.2	2.7	5.2	1.2
1960	249.5	170.8	1.3	4.3	61.0	12.1
1970	548.9	265.4	1.6	5.4	198.3	78.2
1971	581.8	274.6	1.7	5.6	211.1	88.8

* Area within contemporary U.S.S.R. frontiers.
Source: *Transport i svyaz'SSSR*, 1972, 19.

Figure 18 shows graphically the steep rise in the total road passengers carried by public services and, although the curve has risen rather less steeply in recent years, it seems highly probable that, in a few years, it will intersect that of the more gradually rising rail movement, which has also levelled out again after the steep rise of 1965–8.

Bus service statistics are divided into inter-city, suburban and urban. Between 1960 and 1970 the number of inter-city

---

[1] *Transport i svyaz'*, 1972, 99–103.

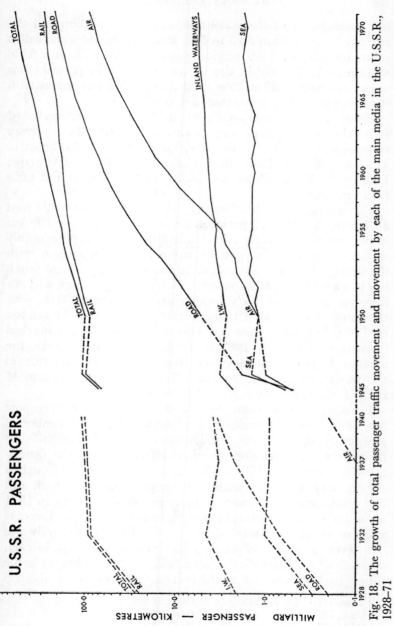

Fig. 18. The growth of total passenger traffic movement and movement by each of the main media in the U.S.S.R., 1928–71

bus passengers rose from 519 million to 1469 million and the average journey from 30·5 to 33·2 km.[1] Bus services classed as suburban carried 1550 million passengers in 1960 and 5378 million in 1970. Within the urban areas the increase in bus usage was from 9247 million to 19 518 million passengers with the average journey rising from 3·6 to 4·8 km.[2]

Other major contributors to urban transport included (with the number of passengers carried in 1970) the tramway systems featured in 110 towns (7962 million), trolleybuses in 111 towns (6122 million), the underground railways of Moscow, Leningrad, Kiev, Tbilisi and Baku (2294 million)[3] and taxis (1144 million).[4]

The average length of journey fluctuated between 34 and 39 km in the 1960–70 period, after a drop from over 100 km in the twenties and early thirties and 50 km in 1950, presumably as suburban services were developed.[5] By comparison with the thirties, however, the role of inland waterways is now much less in total passenger movement. Carriage of passengers by sea also has shown relatively modest growth since 1950–1, with the number embarking rising yearly to a peak of 41 million in 1969, but with passenger-kilometres showing no marked increase except from 1965 to 1968, so that, in general, the average length of journey has decreased. In the late thirties this was about 275 km but had fallen to 50 in 1960 and to 44 in 1971, with even lower figures in some years.[6]

Even more rapid than the rise in road passenger transport has been that in passenger air transport, both in number of journeys and in passenger-kilometres recorded. Insignificant before the war, aviation acquired its position as the fastest-growing carrier in 1957 and has retained its lead. Although it affects only a small number of people compared with rail and road, nearly five times as many people were flying in 1971 as in 1960 and, in terms of passenger-kilometres, the increase was greater and the total about one-third that of the railways. The average length of journey has increased slightly in each successive year since 1962 (after a fall from higher figures in the fifties) to 1201 km, but as the increase has been only about 3

---

[1] *Tran. i svyaz'*, 1972, 242.   [2] Ibid., 244–5.
[3] Ibid., 257.   [4] Ibid., 247.
[5] Ibid., 175.   [6] Ibid., 139.

per cent per annum it has not been responsible for more than a little of the passenger-kilometres increase.

It should also be noted that these figures include the international operations of Aeroflot but these comprised in 1971 only 1·2 per cent of passengers and 3 per cent of passenger-kilometres, so that their inclusion does not significantly distort the pattern if it is thought of in terms of transport in the Soviet Union.

The growth of freight and passenger movement having been examined, it is now appropriate to consider whether the transport available in the U.S.S.R. appears to meet the needs of the economy and the population. To examine this question thoroughly would need a detailed study but some preliminary impressions can be advanced.

There can be no doubt that the transport system of the U.S.S.R., after the rapid development portrayed in the foregoing statistics and in the previous chapters, is much more able to cope with its task than it was even a decade ago while, compared with the Stalinist era, it is infinitely better equipped and organised. There is, however, ample evidence that the system and its constituent parts are not wholly adequate to the demands placed upon them and from time to time suffer severe strain.

The railways, with their routes (for public or general use) totalling 135 200 km and claiming 2637 milliard ton-km of freight traffic and 582 milliard passenger-km, are undoubtedly the most heavily used major system in the world. While this is good for their economics it means that they can have little spare capacity and delays are common. The Soviet press does not comment officially on their shortcomings but there are frequent complaints from users, particularly in agricultural areas, of failure to move goods including crops and fertilisers (which are liable to severe deterioration) to their destinations promptly. Similarly, it is only necessary to visit a railway station in Moscow during the holiday season to see the enormous crowds that have to struggle for a place on long-distance trains.

The extension of lines, double-tracking and other improvements planned for Soviet railways have been remarked in Chapter 3, but they still have a long way to go in technical improvements. A high dependence on manual labour, including that of women, even on track maintenance, is still a feature obvious to all travellers and increased mechanisation is probably essential to speed up trains and achieve still better track utilisation.

The unsatisfactory state of road and road vehicle development in the Soviet Union is well known. Despite the difficulties of terrain and climate, more and better roads are urgently needed and construction is being pushed ahead, but a modern road system seems far away. The official figure of 511 600 km of hard-surfaced roads probably overstates the present position because many of the 'hard' surfaces are rough and easily damaged by the traffic of heavy lorries to which they are subjected. Furthermore, many rural roads have only a narrow carriageway of bitumen and passing vehicles have to swerve off continually on to unsealed side strips, a practice which is dangerous and tiring for drivers as well as destructive of vehicles. Although private ownership of cars is still at a low level, traffic jams are increasingly common in the larger cities and in smaller towns which have grown rapidly with industrial development. On the other hand, the authorities have made energetic efforts to improve traffic control with special police and traffic signals and, as the bulk of passenger movement is by public services, the appalling congestion and waste of effort characteristic of western cities is not, as yet, paralleled in the U.S.S.R.

The production priority for trucks and public service vehicles in the motor vehicle industry is not sufficient to guarantee replacement of aged vehicles, nor is the output and distribution of spares adequate to achieve the high level of reliability of operation which is the aim of the planners. At the same time, it must be granted that the achievement of widespread regular and extremely cheap passenger services in difficult conditions represents no small success for Soviet planning.

There is also no denying the achievements of Soviet airlines, which, in 1971, logged nearly ten milliard ton-km of operations (passenger, freight and mail), all but 420 million being inside the U.S.S.R. This, like the 88 milliard passenger-km performed (all but some three milliard domestically), was the second highest total in the world. The U.S.A., however, recorded the much higher totals of over 21 milliard ton-km and 179 milliard passenger-km on internal scheduled services.[1] In addition, the U.S.A. has a high rate also of charter and private business flights. Aeroflot, of course, achieves a very high load factor—up to 80 per cent for freight and about 75 per cent for passengers on annual overall figures. Indeed, it is rare to find even a single seat unoccupied on many Soviet domestic flights, but this, though economically attractive, indicates the pressure which exists on the system which appears to have little spare capacity. The domestic network of 596 000 km, excluding overlapping sectors, with over 3000 sectors listed in the central timetable alone, is certainly one of the most impressive in the world, but development probably needs to be pushed ahead yet more rapidly to meet the ever-growing demand.

Of recent development, present capacity and future projects in relation to requirements in the pipeline, inland waterway and domestic shipping sectors, little can be said because of the lack of statistics and written comment from within the U.S.S.R. Canal systems have so improved the inland waterway link routes that they must be a valuable addition to rail routes but they suffer from difficulty in planning fully effective utilisation owing to the seasonal interruption of services. It is reported that 70 per cent of most dry goods and 30 per cent of fuel oil shipped on the inland waterways may take two seasons to reach their destination.[2] To explain this very high figure it would seem that the northern sea route and marine shipments from west to east, etc. must be included in this estimate.

The importance of the developments in the merchant marine, as noted in Chapter 4, are mainly international, but the Caspian and coastal services have appreciable value locally for industrial traffic.

In a command economy transport provision can be at a lower level than is expected in capitalist societies because of the absence of directly competing services. Thus, if there is a railway between two points, the manufacturer can be required

---

[1] I.C.A.O. statistics, *Flight*, 28 June 1973, 976.
[2] *Flight*, 24 May 1973, 769.

to use it; he will not be permitted to ignore its existence and send his goods by road simply because he can get a cheaper haul and ignore any social costs that arise from the use of the road transport in preference to rail. Similarly, in the command economy there is no place for the one car–one occupant component of the journey to work, now common in capitalist societies and often accompanied by nearly empty public buses and trains. Hence, the mere fact that there are fewer transport services in relation to the level of industrial activity, the number of population or the area of the country in the U.S.S.R. compared with North America or western European countries does not necessarily mean that the quality of life is correspondingly lower. Even if it is lower at present, it is possible that it will not always be so if the Soviet Union can make its transport system as comprehensive and unified as it plans, while capitalist countries become increasingly congested with duplicated and substantially unnecessary traffic movements. The freedom of choice preferred in western society may become illusory because of costs, both monetary and social, which appear to threaten severely the future of town and country alike.

Because it is later in development than many countries, the U.S.S.R. can learn from the experience of those who are suffering the almost uncontrolled congestion in cities caused by private cars and lorries and the unrestricted invasion of beautiful and agricultural countryside by motorists from the cities. On the other hand, the capitalist countries concerned with moderating the effect of over-rapid development of private transport can learn from the U.S.S.R. the merits of a planned public transport system, charging low fares for an effective, if not luxurious, service. The subsidisation of such services need not be excessive if utilisation is high and may, in any case, be justifiable on social grounds.

At the time of writing, however, it does seem that the Soviet Union is still somewhat undercapitalised in its transport network, and that the physical capabilities of this network are not wholly adequate for the needs of the nation.

Some Important Dates in the History of Transport in Russia and the U.S.S.R.[1]

1580–4	Construction of the port of Arkhangelsk
1648	Dezhnev's navigation of Bering Strait
1697	First Russian trading caravan to China
1703	Founding of St Petersburg and commencement of Vyshnevolotskaya canal system, and development of Kronstadt port
1716–17	Kozma Sokolov, the Cossack, sailed across the northern part of the Okhotsk Sea to open the sea route to Kamchatka
1725–9	Bering's first Kamchatka Expedition
1733–43	Second Kamchatka Expedition (Bering and Chirikov) also known as Second Academy Expedition
1753	Decree on the abolition of internal customs duties
1773	Completion of Yakutsk Track from Irkutsk to Yakutsk along the R. Lena, linking with the Yakutsk-Okhotsk Track
1794	Odessa founded, port opened 1795
1797	Post of Superintendent of Waterways created
1798	Department of Waterways set up
	Main Directorate of Ways of Communication set up
1810	Mariinskiy canal system opened
1811	Tikhvinskiy canal system opened
1815	Construction of the first steamship in Russia
1834	St Petersburg–Moscow highway surfaced
1837	Opening of the first railway line in Russia: Tsarskoye Selo–St Petersburg
1842	Railway Department created

[1] Dates selected are of events important in themselves or referred to in the text.

1849	Steam vessels introduced on the Volga
1851	Opening of the Nikolayevskaya (now Oktyabrskaya) railway, St Petersburg–Moscow
1860	Port of Vladivostok founded
1862	Moscow–Nizhniy Novgorod railway completed
1864	Moscow–Ryazan railway completed
1865	Ministry of Transport (*Putey soobshcheniya*) set up
1866	Ryazan–Kozlov railway completed
1867	Ryazhsk–Morshansk railway completed
1868	Moscow–Kursk railway completed
1869	Kozlov–Voronezh railway completed
1870	Moscow–Yaroslavl, Moscow–Smolensk and Rybinsk–Bologoye railways completed
1871	Kozlov–Saratov railway completed
1879	First oil pipeline, Balakhany–Baku (9 km)
1878–82	Trans-Ural railway extended from Perm and Sverdlovsk to Tyumen
1880–1900	Transcaspian (after 1889 Central Asian) Railway under construction
1884	Yekaterinin railway completed
1892	First electric tramway in Russia constructed in Kiev
1892–1901	Trans-Siberian railway under construction
1900–6	Orenburg–Tashkent railway under construction
1913–31	Turkestan–Siberian railway under construction
1917	10 Nov. (28 Oct.) Decree creating Bureau of Commissars of aviation and aerostation, the organ directing the formation of the Red Air Force (*Krasnyy vodushnyy flot*)
1918	28 June, Decree of the Council of People's Commissars on the nationalisation of all large scale industry and rail transport 4 Sept., Decree on the liquidation of privately owned railways
1922	1 May, Joint Soviet-German airline Moscow–Königsberg 25 Aug., Moscow–Nizhniy Novgorod service instituted under the special organisation *Aviakul'tura* serving the Nizhniy Novgorod fair with the aircraft *Il'ya Muromets*
1923	9 Feb., Resolution of the Soviet of Labour and Defence to create Soviet of Civil Aviation (*Sovet grazhdanskoy aviatsii*) April, creation of *Dobrolet* (*Rossiyskoye obshchestvo dobrovol'nogo vozdushnogo flota*), *Ukrvozdukhput* (*Ukrainskoye obshchestvo vozdushnykh soobshcheniy*), *Zakaviya* (*Zakavkazskoye obshchestvo*). Regular service Moscow–Nizhniy Novgorod started
1923–5	Establishment of personnel in the airline organisations
1924	Moscow–Nizhniy Novgorod air service extended to Kazan First diesel railway locomotive built in U.S.S.R.
1926	First Soviet electric railway Baku–Surakhany (20 km) Kazan–Sverdlovsk railway completed
1927	Petropavlovsk–Borovoye, Gorkiy–Kotelnich and Dnepropetrovsk–Kharkov railway lines completed
1930	1 May, Turkestan–Siberian (Turksib) Railway opened

SOME IMPORTANT DATES

1931	Reconstructed A.M.O. (now *Likhachev*) motor plant, Moscow, opened.
	Borovoye–Karaganda railway opened
1932	Navigation of Dnieper rapids made possible by completion of Dneproges dam.
	First north–east passage in one season made by 1400-ton *Sibiriakov*.
	Aeroflot created to unify all Soviet civil air services.
	Gorkiy motor car plant opened
1933	Moscow–Irkutsk air service commenced.
	20 June, opening of the White Sea–Baltic Canal.
	First use of trolley-buses in U.S.S.R.
1933–4	Polar expedition by the icebreaker *Chelyuskin*.
	First freighters convoyed on Northern Sea Route
1935	15 May, first stage of Moscow Underground Railway opened
	Urals Wagon-Building plant (Nizhniy Tagil) opened.
1937	15 July, opening of the Moscow Canal
1938	Opening of the second stage of the Moscow Underground
1939	Opening of the Karaganda–Balkhash, Uralsk–Iletsk and Lokot–Leninogorsk railway lines
1940	Opening of railway lines, Zharyk–Dzhezkazgan, Volochayevka–Komsomolsk-on-Amur and Ulan Ude–Naushki
1941	Moscow Canal completed
1946	Opening of railway Komsomolsk-on-Amur–Sovetskaya Gavan, followed in 1954 by opening of line Tayshet–Ust-Kut, forming parts of major Baykal–Amur magistral proposed in 1938 and commenced in 1974
1950	Completion of railway Kotlas–Pechora coalfield
1952	First merchant ships with a capacity in excess of 10 000-tons deadweight acquired (13 250-ton motor tankers built in Denmark)
	Volga–Don canal completed
1953	First post-Second World War merchant ship to be completed in U.S.S.R. (16 000-ton tanker)
1956	First post-Second World War dry cargo merchant ship built
	First Soviet jet airliner (Tu-104) in service
1958	First post-Second World War ocean-going passenger ship (*Mikhail Kalinin*) completed in East Germany.
	First mother ship completed in Poland for Soviet fishing fleet
1958	Pipeline for oil from Volga area to refineries at Perm and Gorkiy opened
1960	Nuclear powered ice-breaker *Lenin* in service (16 000 tons)
1961	Direct non-stop air service Moscow–Khabarovsk inaugurated with 220-seat Tu-114
1962	Pipeline for oil distribution Omsk–Angarsk opened

1964	Friendship oil pipeline to Eastern Europe open to Poland, East Germany, Czechoslovakia and Hungary Opening of Volga–Baltic canal Passenger cruise liner *Ivan Franko* commissioned
1968	31 Dec., first test flight of supersonic airliner, Tu-144 (achieved first supersonic flight by any airliner on 5 June 1969)

Index

Accra, 148
Admiral Makarov (icebreaker), 134
Admiral Scheer (German cruiser), 138
Aeroflot, 99, 142–63, 165–77, 180–1
Afghanistan, 49, 55; frontier of, 50, 54, 55
Africa: airlines, 144, 151; trade, 111, 112, 114
agriculture, orientation of transport to, xvii, 1–74, 79, 90, 103
aircraft design, 143
airports, 143, 157–63; classification, 157–62; runways, 159–62
airships, 99
air transport, 99, 142–63, 165, 180–1; freight traffic, 84–5, 86–7, 99, 102, 166–70, 177; passenger traffic, 84, 86–7, 99, 171–7; length of haul, 87; role, 88, 99, 102–4
AK-1 (aircraft), 144
Akhal-Tekke oasis, 47
Alaska, 140, 141
Aldan river, 140
Aldan highway, 98, 140
Aldan region, 95
Aleksandr Pushkin (vessel), 114
Aleksandrov-Gay, 55
Alma-Ata, 150
Altay region, 52, 53, 54; railway, 53
America: aviation, 152, 157; grain sales, 117; ships, 108, 117, 122; technology, 91, 99; trade, 111, 112, 117, 118, 124, 126; *see also* U.S.A.; America, Latin
America, Latin/South, 111, 112, 118 123, 152

Amsterdam, 154, 160
Amu Dar'ya: region, 55, 62, 95; river, 49, 51, 56, 57
Amur: region, 95, 98; river, 90, 95
Amur–Baykal railway, 181
Amur–Yakutsk highway, 140
Anabar river, 132, 140
Anadyr', 95
Andizhan, 49, 54, 57, 58, 61, 63, 65
Angara–Ilim region, railways, 94
Angarsk, 181
Anglo-Russian friction in Central Asia, 46, 49
Anglo-Soviet Shipping Co., 123
Annenkov, General M. N., 47
Antonov aircraft, 143, 149–50, 162
Aral Sea (Aral'skoye More), xvi, 51, 55
Arctic and Antarctic Research Institute, 141
Arctic Ocean, xvi, 102, 113, 129–41
Arctic region, xvi, 151
Arkhangel'sk (Archangel), 127, 135, 150, 179
Arktika (icebreaker), 134
Armenia, 94, 150, 155
Arys', 53
Ashkhabad, 57, 151
Astrakhan', 36, 95; tract, 30
Aul, 53
Australia: airlines, 152; trade, 116, 118, 123, 124–5
Aviakul'tura, 180
Aviation, Soviet of Civil, 180
Azerbaydzhan, 150, 155
Azov, Sea of, 7fn, 100

Bakayev, V. G. and Bayev, S. M., 113
Baku, 46, 50, 69, 150, 180; underground railway, 174; electric railway, 180
Baku–Batumi railway, 46
Balakhany–Baku pipeline, 180
Balkan trade, 108, 111
Baltic region: air travel, 155; buses, 98
Baltic Sea and ports, 26, 28, 34, 80, 100, 102, 106, 111–14, 119
Baltic States, annexation of, 108
Baltic Steamship Company, 117, 118, 125
Baltic–Volga canal, 182
Baranskiy, N. N., 76
Baranov commission, 47
Barents Sea, 127
Barnaul, 53, 153
Batory (vessel), 110
Batumi, 50, 69
Bayev, S. M., 113
Baykalia: 83, 88, 103; railways, 95, 181; roads, 98
Baykov, A., xx
Beeching Report (British Railways), 93
Belgium, 109
Beloozerskiy canal, 36
Belorussia, 91, 94, 150, 155
Belov, M. I., 127
Beloye, lake, 36
Bering, V., 179
Bering Strait, 127, 179
Berlin, 144, 151
Biysk, 53
Black Sea and ports, xv, xvi, 7fn, 46, 50, 80, 83, 84, 99–100, 106, 111–14, 116, 119, 124
Black Sea coast, airlines, 162
Bliokh, I. S., 9, 11
Bogoroditsk, 30
Bologoye, 180
Bolotnaya Square, 33
Bol'shoy Never, 95
Borisoglebsk, 32, 33
Borkovskiy, I., 11, 20, 26
Borok landings, 32, 33
Borovoye–Karaganda railway, 181
Bosporus, 119, 122
Bratislava, 151
Bratsk, 95
Brazil, trade, 116, 118
bridges, *see* engineering problems

British: aircraft and airlines, 144, 147, 149; influence in Central Asia, 46; ports, 125; railways, 93; shipping, 106, 137, 138; *see also* United Kingdom
Bruickmeier, A. O., 55
Budapest, 151
Bukhara, protectorate of, xvii, 46, 47, 49, 51, 54, 55, 71; cotton, 54–5, 57, 60, 62, 71; railway, 54, 62
bulk carriers, 108, 119, 122–3, 125
Bulgaria, 112; trade, 109, 115
Bureau of Commissars of aviation and aerostation, 180
buses, *see* road transport
Buzuluk, 70
Byt Gor, 41

Canada: icebreaking, 134; legislation, 137; trade, 103, 111, 114, 116, 124
canals, *see* waterways, and individual canals
Caspian Sea, xvi, 83, 84, 90, 99–100, 177; railways, 46, 47, 50, 61, 94, 95; *see also* Transcaspia
Caucasus regions: airlines, 151, 162; buses and roads, 98, 100; railways, 47, 50, 83, 88, 94; sea transport to, 83
Central Asia (Soviet), 78, 80; airways, 102, 147, 159–60; buses and roads, 98; railways, 90, 94, 100; underdevelopment, 78, 80
Central Asian Railway, 54, 55, 64, 65, 67, 69, 70, 180; cotton shipments, 57–62; extension to Fergana, 49; impact on caravan route, 50, 56–7, 58–9; impact of Tashkent Railway, 59–60; impact on trade routes, 49–50
Central Blackearth region, 83, 97
Central Industrial region, 83, 97
Central Statistical Committee, 21
Ceylon, 119
Chardzhou, 55, 62
Chelyabinsk, 153
Chelyuskin (icebreaker), 181
Chernigov province, 23
Chile, trade, 117, 118
China: strategic, 95; trade, 68, 109–11, 113, 114, 116, 123, 179
Chirikov (and Bering) expedition, 179
Chukotka (Chukotskiy N. O.), 131

INDEX

Chul'man coalfield, 95
climate, xvi–xvii, 80, 89–90; *see also* winter, permafrost
coal, 79, 84 (flow map), 92, 93, 95, 103, 110; use by shipping, 108
Comecon, 103, 109–12, 118, 121
commuting traffic, *see* urban/suburban transport
container traffic, 96, 103, 119, 122–5, 135–6
Copenhagen, 151
corduroy roads, xvi, 98
cotton (19th cent.), American imports, 64; area, 65, 70–1, 72; impact of war, 72–3; imports from Afghanistan and Persia, 59–60; obstacles to development of cultivation, 63–5, 71–2; (1970) length of haul, 92
Cotton Committee, 63, 68, 69, 71
Council of People's Commissars, 180
Crimean War, xix, 13, 29
Cuba, 112–14, 116, 119
Czechoslovakia: aircraft, 146, 149, 162 164; trade and shipping, 103, 109, 111, 112, 116

Dankov, 30
Danish Sound, 122
Danish Straits, 119
Danube ports, 116, 124
Delhi, 148
Denmark, 109, 181
Department of Waterways, 179
Dezhnev, S., 179
Dikson, 132
distance as operating factor, xv, xvi–xvii, 91–2, 153
Dneproges dam, 181
Dnepropetrovsk-Khar'kov railway, 180
Dnieper (Dnepr) railway, 93; river system, xvi, 83, 97, 100, 181
Dobrolet, 143, 180
Donbas, xviii
Donets railway directorate, 93
Don river, xvi
Douglas aircraft, 146, 161–2
Dudinka, 131, 135
Dushanbe, 151
Dutch aircraft, 149; shipping, 122; *see also* Netherlands
Dzhalal-Abad, 54, 62
Dzhezkazgan, 181

East Europe, merchant fleets and trade, 106–26 *passim*; *see also* Comecon
East Siberian Sea, 130, 134
European Economic Community, comparison, 125; trade, 123
Egypt, trade, 114
Emancipation of serfs, 3, 4, 13, 26
Emba oilfield, 55, 95
Emba river, 55
engineering problems, 80, 89–90, 139–40
Estland province, 20
Estonia, 94, 150
exports (19th cent.), 7–9, 12–14, 30fn; (20th cent.), 106–8, 114–26 *passim*

Fairhall, D., 116
Fairplay, 125, 126
Far East, trade, 118; *see also* Far East (Soviet), China, Japan, etc.
Far East (Soviet), 78, 84, 94, 99–100, 103, 106–7, 110, 113, 116, 123, 127, 128 (map), 150–4; *see also* Pacific Ocean
Far East Maritime Shipping Co., Soviet (FESCO), 120, 123
Fedorov, N. P., 69, 70
Fergana: railway, 54, 62; oblast, 61, 68, 71, 72; valley, xvii, 49, 50, 54, 56, 62, 63, 64, 65, 66, 67, 70
ferries, 81 (map), 90, 123
fertilisers, transport of, 92, 123
Filonovo, 33
Finland, 28, 108, 109, 124, 133, 134
fisheries, 112, 113–14, 138, 181
Fishlow, A., xxi
fish, transport of, 92, 109, 113–14
Five-Year Plans, xxiii, 78, 93, 95, 107, 111
Fogel, R. W., xxi
foreign aircraft, 165; finance, 53; ships 108; technology, 91, 99, 143
freight transport: air 86–7, 102, 146, 153, 165–70; flow maps, 81, 82, 84, 85; inland waterways, 86–7, 166–74, 177; length of haul, 85, 87; pipeline, 86–7, 166, 169–70; rail, 80 *et seq.*, tables, 86–7, 165–70; road, 80 *et seq.*, tables, 86–7, 165–70; sea, 86–7, 99, 100–26, 131–41, 165–74, 177; volume, 79, 83, 86–7 (tables), 88; *see also* cotton, grain
France, 114, 150

Fokker aircraft, 143, 149
Frunze, 150
fuel supplies, 78–9, 89–91, 103, 131

gas, natural; traffic, 103, 166
Gdynia, 122
Georgia, 94, 150, 155
Germany, 99, 108, 138, 143
Germany, East, 109–12, 114, 146, 181
Germany, West, 114
Gibraltar, 125
Glavnoye Upravleniye Zemleustroystva i Zemledeliya, 68, 71
Glavsevmorput', 136–7
GOELRO, 90
Goldmerstein, L. M., 68
goods traffic, *see* freight
Gor'kiy, 143, 181
Gor'kiy–Kotel'nich railway, 180
Gosplan, 77
grain (19th and early 20th cents.): consumption, 4, 5fn, 10fn, 26–30; domestic retentions, 13–14; exports, 7–9, 12–14, 30fn; harvests, 7–8; movements, 5–6; provincial yields, 17–19; sowings, 17–19, 24; supply to Moscow, 30–4; to St Petersburg, 34–9; to Turkestan from Semirechiye, 63, 68, 71, 73; by Central Asian Railway, 65–6, 68; by Tashkent Railway, 67–70, 71–2
grain surplus, potential: definition of, 4–5; provincial breakdown, 17–19; release of, 9–14; total level, 9
grain transport (20th cent.): length of haul, 92; sources of supply, 103, 106; traffic flow map, 85; U.S. sales, 117
Greeks, xvi
Greek shipping, 122
Greenland Sea, 130
Gryf organisation, 112
Gubarevich–Radobyl'skiy, A., 63
Gyatsintov, N. N., 55

Hansa (vessel), 114
Harbron, J. D., 107, 108, 110
Havana, 148
Hawke, G. R., xxi
Hong Kong, 123
Hopemount (British vessel) 138
Hungary, trade, 109, 111, 112, 115–16
Hunter, H., xx

ice, inland waterways, xvi, 90, 97; northern sea route, 129–30, 132–41
icebreaker vessels, 132, 133–41
ice, methods of breaking, 134
Igarka, 130, 135, 137
Il'men', lake, 37, 94
Il'ya Muromets (aircraft), 180
Ilyushin aircraft, 143, 146–50, 161–2
imports, 106, 108–26
India, trade and communications, 111, 112, 114, 116, 118, 119, 144, 148
Indigirka river, 132, 139
industrial complexes and transport demand, xvii, 77–9
inland waterways, *see* waterways
Intourist, facilities at airports, 161
investments in transport, xix–xxiii, 53, 107, 119, 140, 164–5, 177–8
Iraq, 123
Irkutsk, 179; airlines, 144, 150 153, 181; airport, 161
iron ore, *see* minerals
Irtysh river, 139
Italy, trade, 108, 114
Ivan Franko-class vessels, 114, 124, 182
Ivanovo, xvii

Japan, 103; trade, 108, 111, 114, 116, 123, 125; strategic aspects, 95, 133

Kagan, 54
Kalinin-class vessels, 114
Kaluga province, 19, 23, 24fn
Kama region, 6fn
Kama river, 36
Kamchatka, 179
Kapitan-class icebreakers, 133
Karaganda, 79, 93–4, 181
Karaganda–Balkhash railway, 181
Kara Sea, 127, 131, 138
Kara-Kum desert, 47
Karshi, 54
Kashgar, 63
Kashira, 30
Kaspiysko–Aral'skaya railway, 55
Kaufmann, General, 56
Kazakhstan, 94, 95, 150
Kazalinsk, 50, 56
Kazan', 144, 180; province, 23, 30
Kazan'–Sverdlovsk railway, 180
Kazanskiy, N. N., 76–7, 81, 82 (maps), 83, 97

INDEX

Kerch', straits of, 90
Khabarovsk, 148, 150, 153
Khandyga river, 140
Khar'kov, 180
Khatanga river, 132, 140
Kherson province, 20
Khiva, protectorate of, 46, 47, 51, 55, 57, 60, 62, 71
Khurasan, Persian, 49, 57
Khrushchev, N. S., 102
Kiev, 106, 180; airlines, 151, 153; airports, 159–60; underground railway, 174; province, 20
Kievan confederation, xvi
Kinel', 51, 67, 70
Kirgizia, 94, 150
Kirgizstan-class vessels, 114
Kirov, 93
Kishinev, 151
Kitab, 54
Kizel–Sverdlovsk line, 90
Kizyl-Arvat, 46, 47, 49
Kokand, 54
Kokan-Kishlak, 54
Kolomna, 30, 32, 33, 42, 180
Kolyma river, 131, 135, 140
Komet (German vessel), 138
Komi A.S.S.R., 150
Komsomolsk (Komsomol'sk-na-Amure), 90, 95, 181
Königsberg, 143, 180
Kopet-Dag mountains, 47
Korea, North, trade, 110, 116, 123
Kostroma province, 23, 24fn
Kotel'nich, 180
Kotlas–Pechora railway, 181
Koval'chenko, I. D., 22
Koval'evskiy, V. P., 54
Kovzha river, 36
Kozlov, 30, 32, 33
Kozlov–Saratov railway, 180
Kozlov–Voronezh railway, 180
Krasnovodsk, 47, 49, 57, 59, 60, 61, 65, 66, 68
Krasnoyarsk, 132, 144, 150
Krivoshchein, A. V., 63, 71
Krivoy Rog, xviii, 91
Kronstadt, 30, 37, 179
Krypton, C., 127
Kugart pass, 63
Kungrad, 55, 95
Kuropatkin, General A. N., 53
Kursk, 33, 180

Kursk province, 5fn, 10fn, 12fn, 22, 23, 25, 30, 32, 33, 34, 180
Kushka, 50
Kuybyshev, 151
Kuzbas, xviii, 78–9

Ladoga, lake, 36
Laptev Sea, 130, 134, 138
Latvia, 150
Lavrishchev, A., 75–6
Lazarenko, T., 107
Lebedyan', 30
legal aspects of navigation, 137–8
Lena region: railways, 94, 95, 139; roads, 140; sea access, 135
Lena river, 89, 131–2, 179
Lenin, V. I., 1
Lenin (icebreaker), 133–4, 181
Leningrad, 80, 123; airlines, 150, 151; airport, 159–60; shipbuilding, 114, 122; underground railway, 174; *see also* St Petersburg
Leninogorsk, 181
Let L-410, 149, 160
Levant trade, 112
Liberia, 125
Liberty (U.S.) ships, 108
Lifland, 20
Likhachev (formerly A.M.O.) motor plant, 181
Linkage effects of railway construction, xxi
Lisunov Li-2 aircraft, 146, 149
Lithuania, 94, 150
Livna, 30
locomotives, railway, 89–91, 180
Lokot'–Leninogorsk Railway, 181
London: airlines, 151, 160; shipping, 123
London Stock Exchange, 54
Lovat river, xvi

Magadan, 139, 140, 150
Main Directorate of Ways of Communication, 179
Maklakovo, 95
Manchuria, 95
Manhattan (supertanker), 141
Mankent, 53
Mariinskiy canal, 36
Mariinskiy canal system, 36, 39, 41, 179
Maxim Gor'kiy (aircraft), 144

Mediterranean trade, 109, 123
merchant marine, *see* shipping
Merv, 50
Merv Turkmens, 49
Mikhail Kalinin (vessel), 181
Mikhail Lermontov (vessel), 125
Mikhaylov, 30, 32
Mikhaylovskiy Zaliv, 47
minerals, xvii–xviii, 79, 92–3, 110, 117, 118, 123, 131; *see also* coal, petroleum
Mineral'nyye Vody, 144
Ministry of Civil Aviation, 157
Ministry of Communications (or Transport), 49, 59, 180
Ministry of Finance, 68, 71
Ministry of Foreign Affairs, 68
Ministry of War, 53
Minsk, 150; province, 23
'Mixed Corporations', 108, 111
Mogilev province, 23
Moksha river, 32
Moldavia, 94, 151
Mologa river, 37
Monchegorsk, 131
Morshansk, 32, 33, 41fn, 42
Montreal, 148
Morskoy-21-class vessel, 135
Morskoy Flot, 129
Moscow (Moskva), xvii, xviii; airlines, 99, 143–63; airports, 159–62; container transport, 123–4; grain supply (19th cent.), 26–8; historical aspects of grain market, 2, 6fn, 7fn, 11, 24fn, 26, 28, 30, 32, 33, 34, 37, 41fn, 43; suburban transport, 90, 174, 181
Moscow Canal, 181
Moscow duma, 33
Moscow province, 16, 23, 24, 24fn, 25, 28
Moscow Radio, 129
Moscow region, 61; cotton imports, 47; cotton manufactures, 70
Moscow–Kolomna railway, 33
Moscow–Kursk railway, 33, 180
Moscow–Nizhniy Novgorod railway, 33, 180
Moscow–Ryazan' railway, 32, 33, 34
Moscow–Ryazan'–Kozlov railway, 34
Moscow–Ryazan'–Kozlov–Voronezh railway, 33
Moscow–Smolensk railway, 180
Moscow–Yaroslavl railway, 180
Moskva-class icebreakers, 133

Moskva river, 32, 33
motor vehicles, 102, 164, 176, 178; *see also* road transport
Msta river, 37
Mstino, lake, 37
Mtsensk, 32
Muravka tract, 30
Murgab valley, 50
Murmansk, 127, 131, 132
Muscovy, xvi

Nakhodka, 123, 124
Namangan, 54
Naryn valley, 54, 63
Nayanov, F. V., 132
Netherlands, 109; *see also* Dutch
Neva river, xvi, 30, 36
New Railways Committee, 52, 54, 55, 68, 72
New Russia, 7fn, 21, 25
New York, 148, 150, 160
New Zealand, 118
Nicholas I, xix
Nifontov, A. S., 25fn
Nikolayev, shipbuilding, 122
Nikolayev railway, xix, 24, 33, 34, 37, 41fn, 42, 180
Nikol'skiy, I. V., 80, 84–5
Nizhniye Kresty, 131
Nizhniy Novgorod, 143, 144, 180; province, 23
Nizhniy Tagil, 181
Noril'sk, 95, 131
Northern sea route, 100, 127–41, 177, 181; *see also* Arctic
North Sea trade, 110–11, 112, 114
North Siberian Railway, 94
Norway, 125, 137
Novaya Ladoga, 36, 37
Novgorod province, 23, 24
Novonikolayevsk, 53
Novosibirsk, airlines. 144, 151, 153; airport, 159–60
Novovoronezh (vessel), 139

Ob' river, and railways, 95, 139; sea access, 135
Obshchestvo Vostokovedeniya, 52, 68
Odessa, 6fn, 50, 106, 179
oil, *see* petroleum products
Oka landings, 32, 33
Oka river, xvi, 32, 36, 42
Okhotsk coast and Sea of, 95, 139, 179

INDEX

Oktyabr'skaya railway, 180
Olenek river, 132, 140
Omsk, 70, 181
Onega canal, 36
Onega, lake, 36
Orël, *see* Oryol
Orenburg, 51, 67; province, 23, 25; region, 68, 71
Orenburg railway, 56
Orenburg-Tashkent caravan route, 50, 58, 61
Orenburg-Tashkent railway, 50-1, 52, 55, 180; *see also* Tashkent railway
Oryol (Orël) 32; province, 23, 25, 30, 32, 34
Osh, 63
Ostrov Diksona, 132
Oy Wärtsilä AB, Helsinki, 133

Pacific Ocean, 99-100, 102, 103, 113, 114, 127, 133; *see also* Far East
Pakistan, 110, 111
P.A.K. Ocean Shipping Co. Ltd, 110
Paris, 151
passenger transport: air, 84, 142-63, 165, 171-7; inland waterways, 83, 171-4; rail, 83, 171-6; road, 171-8; sea, 114, 124-5, 171-4; volume, 155, 171-8
Patom region, 95
Pechora, 93, 95, 181
Penza province, 23, 30
Perevles landings, 32, 33
Perm', 180, 181
Persia, 47, 49, 68
Persian Gulf, 49, 119
permafrost, 80, 89, 98
petroleum products: length of haul, 92; orientation of transport to, 46, 79, 94-5; supply, 91; tankers, 113, 119, 122-3, 125; traffic, 84 (flow map), 166-70, 177; use, 103; *see also* pipelines
Petropavlovsk, 114
Petropavlovsk-Borovoye railway, 180
Pevek river, 131, 135
Peter the Great, xix, 36
Petrov, N. P., 62
Petrovsk, 50
Pinkhenson, D. M., 127
pipelines, 81 (map); construction dates, 180-2; role, 88, 102; traffic, 83, 86-7 (tables), 166, 169-70

Pishpek, 53, 63
planning, transport, xv-xxii, 75 *et seq.*, 103; *see also* Five-Year Plans
Polar Record, 137
Polish, merchant marine, 110, 113, 118; shipbuilding, 109-12, 122, 124, 181; trade, 108-12, 115-16
Polish Ocean Lines, 110, 112, 121 (map)
ports, 81 (map), 94, 103, 106-26; northern sea route, 126-41
Prague, 151
Pravda, 129, 137
Problemy Severa, 141
Proliv Vilkitskogo, 137-8
Pron river, 32
Proudhoe Bay, 141
Pskov province, 16, 23
PS-84 aircraft, 146

Railway Battalions, 47
Railway Department, 179
railways: choice of motive power, 89-91; comparison with other media, 96-103, 164-78; construction and operation, 88-93; current developments, 93-6; dates, 178-81; density by province, 17-19; electrification, 90-1, 96; freight traffic, 80 *et seq.*, 86-7 (tables), 92, 165-70; inadequacies and traffic problems, 83, 92-3; industrial, 93; investment, 107, 124, 164, 177-8; length of haul, 78-9, 87, 91-2; links by road and ferries, 81 (map), 90, 123, 127 (map), 139-40; narrow gauge, 94; passenger traffic, 83, 86-7 (tables), 93, 171-8; physical obstacles, 89-90; private, in Turkestan, 51-2, 53-5; rate of construction (19th cent.), 2fn; role and importance, 80 *et seq.*, 94, 100, 164-80; routes, 80 *et seq.*, 81 (map), 131; suburban, 90; traffic density, 88; underground, 174, 181; and regional specialisation (19th cent.), 14-26, 46-74
Red Air Force, 180
Ranenburg, 32
raw materials and transport orientation, 78 *et seq.*, 103
remote areas, development of, 78
Riga, airlines, 150; airport, 159, 162; grain exports, 34

rivers, *see* waterways, and individually
roadless areas, problems of, 97–8
road transport, 97–8; first surfaced highway, 179; freight traffic, 86–7 (tables), 166–70, 175; inadequacies, 98; investment in, 98, 164–5; length of haul, 85, 87–8, 97; northern areas, 139–40; passenger traffic, 83, 86–7 (tables), 97–8, 171–8; role, 88, 100–1; seasonal problems, 90, 98, 139
rolling stock, empty running of, 92–3
Romania, trade, 108–9; 112, 115–16
Rostow, W. W., xxi
R.S.F.S.R.: airlines, 142–63; railway growth, 94–5
Rudoi, Y. and Lazarenko, T., 107
Rum, L. L., 55
Russo-American commercial treaty, 64
Ryazan': airport, 159, 161; landings, 32, 33; railways, 32, 33, 34, 180; province, 6fn, 23, 24fn, 30, 32; tract, 30, 33
Ryazan'–Kozlov railway, 3, 30fn, 180
Ryazhsk, 30, 32
Ryazhsk–Morshansk railway, 33, 180
Rybinsk, 36, 41
Rybinsk–Bologoye railway, 30fn, 37, 42, 180

St Petersburg, xix, 2, 11, 26, 28–30, 34, 36, 37, 38, 41, 42, 179; province, 23, 24, 25, 28
Sakhalin, 123
Salekhard railway project, 95
Samara, 41, 51; province, 23, 25, 30; river, 67; valley, 68
Samarkand, 49, 56, 57, 61, 70, 161; oblast, 60
Sarantsev, P. L., 76
Saratov, 41fn, 55, 149, 180; province, 23, 25, 30, 38, 39; region, 95
Saushkin, Yu. G., 76
Scandinavian shipping, 122, 124
Seatrade, 124
sea transport, *see* shipping
self-sufficiency policy in transport, 77–8
Semipalatinsk, 52, 53, 60
Semirechiye, 52, 54, 63, 68
Semirechiye railway, 53, 63, 67, 72, 73
Sergeyev, Kh. N., 71
Sergiyevskoye, 32

Serpukhov, 32
Sheksna river, 36
Shilovo landings, 32, 33
shipbuilding, 107–11, 113, 115, 119, 122, 125–6, 179, 181
shipping: freight traffic, 86–7, 99–100, 106–26, 131–2, 136, 166–70; ice obstacles, 129–30; length of haul, 84–5, 87, 99; passenger traffic, 87, 114, 124–5, 171–4; role, 83–4, 100–3, 165–81
ship purchases, Soviet, 108, 109–10, 114, 124
ships, *see* individually, and types of vessel
Shmeyn, A., 22
Shota Rustaveli (vessel), 124
Siberia, xviii, 52, 71, 78, 80, 83, 100, 103, 128 (map); air transport, 99, 102, 150–4, 159–62; buses and roads, 98; northern, 129–41; railways, 52, 94–5, 180
Sibiriakov (vessel), north-east passage, 181
Simbirsk province, 23, 30
Siversov canal, 37
skis, use in aviation, 144
Skobelev, 49
Skobelev's campaign, 47
Skopin, 32
Slavin, S. V., 141
Smolensk, 180; province, 23, 24, 25
Sorochinskaya, 70
Social saving, xxi
Sokolov, Kozma, 179
South (industrial) region, 83, 97
Sovetskaya Gavan', 94, 181
Soviet of Civil Aviation, 180
Soviet of Labour and Defence, 180
Soviet Union, area and size, xv, 79; climate, 80; population density, 79–81
Sovietskiy Soyuz (vessel), 114
Sovtorgflot, 106–26
Southampton, 125
Spitsbergen, 130
Stalin, J. V., 111, 146, 174
Stalin-class icebreakers, 133
Stockholm, 151
Stolypin, P. A., xx
Suez Canal, 113, 116, 139
statistics, availability and nature of Soviet, 93, 165

Stavropol', 95
suburban transport, 88, 171–4, 178
sugar, traffic, 106, 116, 118, 119
Summary of World Broadcasts, 129
Sura river, 36
Surakhany, 180
Suram Pass, 90
Svalbard, 130
Sverdlovsk, 90; airways, 144, 151, 153; railways, 93, 180
Svir river, 36
Syas river, 36, 37
Syktyvkar, 150
Syr Dar'ya river, 51; oblast, 62, 68, 70, 71, 72

Tadzhikistan, 94, 151, 155
Tallin, 150
Tambov, 32, 33; province, 22, 23, 25, 30, 32, 34
tankers, 113, 119, 122–3, 125
Tashkent, 49, 57, 58, 61, 67, 70, 180; airlines, 151; airport, 159–60, 161; region, 56, 62
Tashkent railway, 51, 53, 58, 64, 71, 72, 73; impact on cotton cultivation, 59–61; on grain trade, 67–70
taxis, 174
Tarski, I., 76
Tavda, 94
Tayshet railway, 89, 95, 181
Tbilisi, airlines, 150, 153; underground railway, 174
Termez, 54, 55
territorial production complex, 78
Tikhvinka river, 37
Tikhvinskiy canal system, 36–7, 179
Tiksi, 95, 131–2, 138
Tilbury, 125
timber: exports, 106; length of haul, 92; traffic flow map, 85; transport of, 79, 110, 131
Tokyo, 123, 148, 150
Tomsk railway directorate, 93
tourist traffic, 124–5, 132
tramways, 174, 180
Transcaspia, 46–7, 49, 50, 60, 62, 71, 180; *see also* Caspian Sea, Central Asia
Transcaucasus, 64; railways, 50; *see also* Caucasus regions
transport flows, 80–8; maps, 81, 82, 84, 85
transport media (19th cent.), 39–42; relative cost, 40; relative speed, 39–40
transport systems: capacity and inadequacy of, 175–8; classification of movements, 77; dimensional framework of, 79–80; physical setting for, xv, xvi, 80; planning and policy, 75–9, 103
trans-Siberian air route, 153, 161
trans-Siberian container route, 123–4
Trans-Siberian railway: air travel comparison, 153; branch lines, 94–5; construction, 180; container traffic, 103, 123–4; relevance to Central Asia, 52, 69; as road focus, 98; role in access to north, 133, 139, 140; stock, 92; strategic considerations, 95, 133; traffic density, 88
Trans-Ural railway, 180
trolleybuses, 174, 181
Tsarskoye Selo railway, xix, 179
Tselinograd, 153
Tsna river, 32, 33, 37, 42
Tsninskiy canal, 37
Tula, 30, 32; province, 23, 24fn, 30, 34
Tupolev aircraft, 143–4, 147–8, 150, 153, 161–2, 181, 182
Turkestan, 46–74
Turkestan–Siberian railway, 51–4, 68, 69, 70, 180; Arys'–Auliye-Ata section, 53; Arys'–Vernyy, 52; Novonikolayevsk–Semipalatinsk, 52; Semipalatinsk–Vernyy, 52, 54
Turkmens, 47
Turkmenistan, 94, 151, 155
Tver, 36; province, 23, 24, 25
Tvertsa river, 37
Tveretskiy canal, 37
Tyumen', 150, 153, 180
Tyumen'–Surgut railway, 94

U-boats, 138
Ufa province, 23
Ukraine, 94; airlines, 151, 155, 180; buses, 98
Ukrvozdukhput', 180
Ukholovo, 32
Ulan-Ude–Naushki railway, 181
Umet, 33
underground railways, 174, 181
United Kingdom: comparisons, 80, 88, 92–3, 99, 122, 144; trade, 106, 112, 114, 124–5
Ural mountains, xviii, 88, 90, 180

Ural region, 79, 94, 95, 97, 151
Ural–Kuzbas *Kombinat*, 78–9
Urals wagon-building plant, 181
Ural'sk–Iletsk railway, 181
urban transport, 171–4, 177–8
U.S.A.: comparison with, xxi, 88, 98–9, 102, 125, 178; cotton trade, 72; grain sales, 117; ships, 108; trade, 117; vessels in northern sea route, 137–8
U.S.S.R.: area and size, xv, 79; climate, 80; population density, 79–81
Ust'-Kut, 181
Ustyurt plateau, 95
Uzbekistan, 151, 155
Uzun-Ada, 57, 64, 65

Vanino, 123
Varangians, xvi
Vatslav Vorovskiy (vessel), 132
V E B Deutsche Seereederei, 112
Venezuela, 123
Venyev, 30
Vernyy, 52, 54, 68, 70
Vienna, 151
Vietnam, North: trade, 110, 112, 113, 116
Vil'nyus, 150
Vil'son, I., 9, 21
Virgin Lands, 94
Vitebsk province, 23
Vladikavkaz Railway, 50
Vladimir province, 6fn, 23, 24, 25
Vladivostok, 114, 127, 146, 153, 180
Vodnyy Transport, 129
Volga Aeroflot Directorate, 151
Volga provinces, 22, 23, 38
Volga river, xvi, xix, 34, 36, 37, 39, 40, 41fn, 50, 60, 61, 79, 83, 94, 96–7, 100, 180; lower region, 21, 26, 30, 33, 34, 36, 38, 71, 95; upper region, 5fn, 10fn, 25, 36, 38
Volga–Baltic canal, 96, 182
Volga–Don canal, 96, 181
Volgograd, 95
Volkhov river, xvi, 36, 37
Volochayevka–Komsomolsk-on-Amur railway, 181
Vologda province, 24fn
Volynia province, 20
Voronezh, 30, 32, 33, 149, 180; province, 20, 23, 30, 34

Vyshnevolotskaya canal system, xix, 36, 179
Vytegra river, 36

Warsaw, 151
War, Crimean, xvi, xix, 13, 29
War, Second World, 99, 108, 133, 139, 181
waterways, inland, 96–7, 81 (map); construction dates, 179–81; freight traffic, 83, 86–7 (tables), 166–70, 177; grain traffic in 19th cent., xv, xvi, 30, 32–42; ice-strengthened freighters, 135; inadequacies, 79, 96; investment, 107; length of haul, 78, 87; northern region, 135, 140; passenger traffic, 83, 86–7 (tables), 171–4; role, 88, 96–7, 100–2
West Dvina river, xvi
White Sea, 106, 127, 132
White Sea–Baltic canal, 181
winter, effects on transport, xvi, 80; air, 144, 147; inland waterways, xvi, 96, 177; northern sea route, 177; rail 89–90; road transport, xvi, 98
Wismar, 114
Witte, Count S., 107
wool, transport of, 118, 119, 122

Yakovlev aircraft, 143, 149, 162
Yakutsk, 95, 140, 151, 179
Yakutsk–Okhotsk Track, 179
Yana river, 132, 139
Yaroslavl, 180; province, 23, 24, 25
Yatsunskiy, V. K., 21, 24fn, 25fn
Yefremov, 30, 32
Yekaterinin (Catherine) railway, xviii, 180
Yekaterinoslav province, 21
Yelets, 30, 32
Yenisey river: icebreaking, 134; power, 95; sea access, 135–6; traffic to, 131–2, 135; and railways 139
Yepifan, 30, 32
Yerevan, 150
Yermak (icebreaker), 134
Yugoslavia, shipping, 118

Zagorskiy, 71, 72
Zakaviya, 180
Zeravshan valley, 49, 65, 66, 68
Zaraysk, 30
Zharyk–Dzhezkazgan railway, 181
Zurich, 151